The ESSENTIALS of

PROBABILITY

Lutfi A. Lutfiyya, Ph.D.
Associate Professor of Mathematics
Kearney State College, Kearney, Nebraska

REA
Research and Education Association
61 Ethel Road West
Piscataway, New Jersey 08854

THE ESSENTIALS® OF
PROBABILITY

Printed in the United States of America

Library of Congress Catalog Card Number 97-65631

International Standard Book Number 0-87891-840-X

ESSENTIALS is a registered trademark of
Research & Education Association, Piscataway, New Jersey 08854

II-1

WHAT "THE ESSENTIALS" WILL DO FOR YOU

This book is a review and study guide. It is comprehensive and it is concise.

It helps in preparing for exams, in doing homework, and remains a handy reference source at all times.

It condenses the vast amount of detail characteristic of the subject matter and summarizes the **essentials** of the field.

It will thus save hours of study and preparation time.

The book provides quick access to the important facts, principles, theorems, concepts, and equations in the field.

Materials needed for exams can be reviewed in summary form – eliminating the need to read and re-read many pages of textbook and class notes. The summaries will even tend to bring detail to mind that had been previously read or noted.

This "ESSENTIALS" book has been prepared by an expert in the field, and has been carefully reviewed to assure accuracy and maximum usefulness.

Dr. Max Fogiel
Program Director

CONTENTS

CHAPTER 1

BASIC CONCEPTS

1.1 SET THEORY

1.1.1 SET MEMBERSHIP

Any well defined collection of distinct objects is called a set.

The individual objects of a set are called the elements of the set and are said to belong to the set, or to be members of the set. Set membership is denoted by the symbol ε. If A represents the set and x represents an element of the set, then $x \varepsilon A$ is read "the element x belongs to the set A" or "x is an element of A." If x is not an element of A, we write $x \not\varepsilon A$.

We may specify a set by stating in words what its elements are. Another way of specifying a set is to exhibit its element, usually enclosed in braces. Thus, $\{x\}$ indicates the set consisting of the single element x; $\{x, y, z\}$ indicates the set consisting of the three elements, x, y, and z; and if P is the set of all positive integers, by writing

$$K = \{a \mid a \varepsilon P, \text{a is divisible by } 2\},$$

we mean that K consists of all elements, a, having the properties indicated after the vertical bar. Thus,

$$K = \{2, 4, 6, 8, ...\}$$

1.1.2 NULL SET (EMPTY SET)

A null set or empty set is a set with no elements.

If one were to describe a null set by listing the elements, one could use the symbol "{ }." By convention, the symbol ϕ *(hpi)* is used to denote the empty set, so that $\phi = \{ \}$. For example, the set consisting of all college students in the United States who are less than 6 years old is an empty set.

1.1.3 EQUALITY OF SETS

> Two sets A and B are equal if and only if the sets contain the same elements. If two sets A and B are equal, we write $A = B$. If A and B are not equal, we write $A \neq B$.

1.1.4 SUBSETS OF A SET

> A set A is called a subset of a set B if every element of A is also an element of B. A is then said to be "included in" or "contained in" B, and this is denoted by $A \subseteq B$.

If A is a subset of B and B contains at least one element that is not included in A, then A is said to be a proper subset of B. In this case, we write $A \subset B$.

As a consequence of the definition of a subset, we see that

1) Two sets A and B are equal if and only if $A \subseteq B$ and $B \subseteq A$.
2) Since the null set has no elements in it, it is a subset of every set; that is, $\phi \subseteq A$ for all A.
3) Every set A is a subset of itself; that is, $A \subseteq A$ for any set A.

1.1.5 UNIVERSAL SET

In discussing sets, it may be useful to consider an underlying reference set. We shall assume that all sets under discussion are subsets of some fixed universal set symbolized U.

> The universal set, U, is the set that contains all elements being considered in a given discussion.

The universal set may change as our frame of reference changes. For example, in some cases, the universal set may be the set of natural numbers or even the set of all rooms in your house. In section 1.2, it becomes clear that, in probability theory, the sample space S is the reference set.

1.1.6 OPERATIONS ON SETS

In arithmetic, we know that numbers are combined by using such operations as addition and multiplication. Sets can be combined by using the following operations;

1. Union

> Let A and B be two subsets of the universal set U. Then the set whose elements belong to either set A or set B (or both) is the union of A and B and is denoted by $A \cup B$. In symbols,
>
> $$A \cup B = \{x \mid X \varepsilon A \text{ or } X \varepsilon B\}$$

2. Intersection

> Let A and B be two subsets of the universal set U. Then the intersection of sets A and B is the set consisting of all elements which belong to both sets A and B and is denoted by $A \cap B$. In symbols,
>
> $$A \cap B = \{x \mid X \varepsilon A \text{ and } X \varepsilon B\}$$

3. Complement

Let A be any subset of the universal set U. Then the set of all elements of U not in A is the complement of A with respect to U and is denoted by A'. In symbols,

$$A' = \{x \mid X \varepsilon U, X \not\in A\}$$

4. The Difference Between B and A, $B - A$

> Sometimes instead of considering the complement of set A with respect to the universal set U, we are interested in the relative complement of A with respect to a set B. This is often called the difference of B and A; that is, $B - A$. In symbols,
>
> $$B - A = \{x \mid X \varepsilon B, X \not\in A\}.$$

1.1.7 VENN DIAGRAMS

Sometimes, it is helpful to represent sets and set operations by means of Venn diagrams. In Venn diagrams, the universal set, U, is represented by a rectangle and the subsets of U are depicted as various regions of the rectangle. Note that Venn diagrams are not proofs. They only help in visualizing the relationship between sets. For example, the above sets in section 1.1.6 can be represented as follows:

1. $A \cup B$

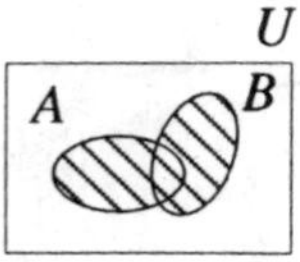

is the shaded region.

2. $A \cap B$

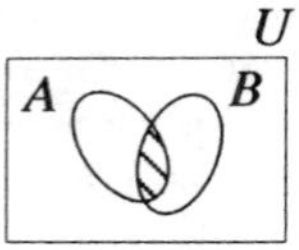

is the shaded area.

3. A'

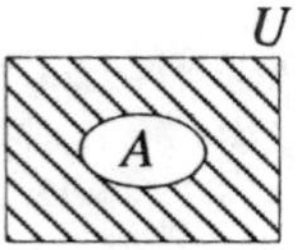

is the shaded area.

4. $B - A$

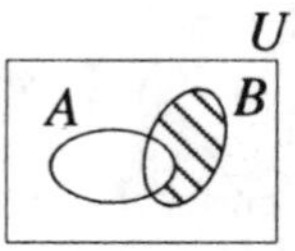

is the shaded region.

1.1.8 FINITE AND COUNTABLE SETS

Sets can be finite or infinite.

A set is finite if the number of elements in the set is zero or a natural number *n*.

For example, the set of letters in the English alphabet is a finite set because it contains exactly 26 elements. Another way to think of this is to say that the set of letters in the English alphabet can be put into a one-to-one correspondence with the set 1, 2, 3, ..., 26.

Infinite Sets

An infinite set is a set that is not finite. That is, a set A can be defined to be infinite if and only if it can be put into one-to-one correspondence with a proper subset of itself.

The set of all natural numbers N is an example of an infinite set.

Countable Sets

A set is countable if it is finite or it can be put in one-to-one correspondence with the set of all natural numbers N, in which case it is said to be countably infinite; otherwise, the set is uncountable.

For example, the set $\{2, 4, 6, \dots\}$ is a countably infinite set.

1.1.9 POWER SETS

Let A be any set. The power set of A, denoted by $P(A)$, is the set of all subsets of A and is written as

$$P(A) = \{B \mid B \subseteq A\}.$$

If A is a finite set having n elements, then $P(A)$ has 2^n elements. For example, if $A = \{a, b, c\}$, then

$$P(A) = \big\{ \{a\}, \{b\}, \{c\}, \{a, b\}, \{a, c\}, \{b, c\}, \{a, b, c\}, \phi \big\}$$

1.1.10 PARTITIONS OF SETS

Let A be a nonempty set. A collection, C, of nonempty subsets of A is called a partition of A if

a) A is the union of all the sets in C. That is

$$\bigcup_{B \in C} B = A$$

b) For any $B, D \in C$, either $B = D$ or $B \cap D = \phi$.

If $A = \{a, b, g, d, e, f\}$, and $B_1 = \{a\ b\}, B_2 = \{g\}, B_3 = \{d, g, f\}$, then $C = \{B_1, B_2, B_3\}$ is a partition of A since,

a) $B_1 \cup B_2 \cup B_3 = A$

b) $B_1 \cap B_2 = \phi, B_1 \cap B_3 = \phi$, and $B_2 \cap B_3 = \phi$.

1.1.11 PRODUCT SETS

> Let A and B be two sets. The product set of A and B is the set $A \times B$ given by
>
> $$A \times B = \{(a, b) \mid a \varepsilon A \text{ and } b \varepsilon B\}.$$

In general, $A \times B \neq B \times A$

Note: Two ordered pairs (a, b) and (c, d) are equal; that is, $(a, b) = (c, d)$, if and only if $a = c$, and $b = d$.

1.1.12 LAWS OF THE ALGEBRA OF SETS

Identity Laws	
1a. $A \cup \phi = A$	1b. $A \cup U = U$
2a. $A \cap \phi = \phi$	2b. $A \cap U = A$
Idempotent Laws	
3a. $A \cup A = A$	3b. $A \cap A = A$
Complement Laws	
4a. $A \cup A' = U$	4b. $A \cap A' = \phi$
5a. $(A')' = A$	5b. $U' = \phi$, and $\phi' = U$
Commutative Laws	
6a. $A \cap B = B \cap A$	6b. $A \cup B = B \cup A$
Associative Laws	
7a. $A \cup (B \cup C) = (A \cup B) \cup C$	7b. $A \cap (B \cap C) = (A \cap B) \cap C$
Distributive Laws	
8a. $A \cup (B \cap C) = (A \cup B) \cap (A \cup C)$	8b. $A \cap (B \cup C) = (A \cap B) \cap (A \cap C)$
DeMorgan's Laws	
9a. $(A \cup B)' = A' \cap B'$	9b. $(A \cap B)' = A' \cup B'$

1.2 SAMPLE SPACES

1.2.1 EXPERIMENTS: DETERMINISTIC OR RANDOM

The word experiment is used to describe any act that can be repeated under given conditions. In the chemistry laboratory, the students determine the boiling point of water as 100° C. If the experimental conditions remain the same, then the determination will be the same.

In contrast, there are experiments in which the results vary in spite of all efforts to keep the experimental conditions the same. For example, coin tossing or the birth date can be thought of as experiments. In these experiments, the results are unpredictable.

> **Definition**: Deterministic and Random Experiments
> If the results of the repeated experiments are exactly the same, we say the experiments are deterministic; otherwise, they are said to be random or stochastic.

Probability theory is used to explain and predict, to some degree, the results of random experiments.

1.2.2 DEFINITION OF A SAMPLE SPACE

> A sample space of a random experiment is the set S consisting of the collection of all possible outcomes of the experiment. The individual outcomes in a sample space are called sample points.

Since we are concerned with the whole collection of all possible outcomes of an experiment, our set is in a sense a full or a universal set of outcomes. For example, if a family of four is selected in some arbitrary way and the total income is calculated to the nearest dollar, then a suitable sample space is the set of all nonnegative integers.

1.2.3 FINITE SAMPLE SPACES

> A sample space is called finite if it contains only finitely many elements or points.

For example, a random experiment of tossing a coin twice is conducted.

If not only the results are important, but also which side comes first and which side comes last, then the four pairs HH, HT, TH, and TT can be used to represent the thinkable outcomes for a sample space. The first letter in each pair represents the result of the first toss; the second letter, the second toss. Thus,

$$S = \{HH, HT, TH, TT\}.$$

Obviously S has 4 elements and hence is a finite sample space.

1.2.4 INFINITE SAMPLE SPACES

> A sample space is infinite if it has an infinite number of elements. A sample space S is countably infinite if the elements of S can be counted; that is, the elements of S can be put in one-to-one correspondence with the set of natural numbers, and noncountably infinite otherwise.

For example, a coin is tossed until the first time a head turns up. Let the outcome of the experiment x, be the first time that a head turns up. Then the possible outcomes of the experiment are

$$S = \{1, 2, 3, \ldots\}.$$

Note that S is a countably infinite sample space.

1.3 EVENTS

1.3.1 OUTCOMES

> An outcome is a particular result of an experiment.

For example, if two coins, one penny and one nickel, are tossed simultaneously and the outcome for each coin is recorded using ordered pair notation: (penny, nickel), then the sample space is given by

$$S = \{(H, H), (H, T), (T, H), (T, T)\}.$$

Note that each of: (H, H), (H, T), (T, H), and (T, T) is a possible outcome of the experiment.

1.3.2 DEFINITION OF EVENTS

> An event is a subset of the sample space. An elementary event is a set consisting of a single element of the sample space.

To say that an event E has occurred is to say that the actual outcome of the random experiment is an outcome in the set of outcomes associated with the event E. Thus, if in the draw of a chip one of the six numbers in the set $\{3, 6, 9, 12, 15, 18\}$ occurs, then we say that the event "the number is a multiple of 3" has occurred; otherwise, the event does not occur.

1.3.3 COMBINATION OF EVENTS

Since sets and events are analogous, events, like sets, may be combined. We can combine events to form new events using the various operations on sets as follows :

1. $A \cup B$ is the event that occurs if and only if A occurs or B occurs or both A and B occur.
2. $A \cap B$ is the event that occurs if and only if A occurs and B occurs.
3. A', the complement of A, is the event that occurs if and only if A does not occur.
4. $A - B$ or $A \cap B'$ is the event that occurs if and only if A occurs but B does not occur.

Every thinkable outcome of the random experiment is represented in the sample space S, so S is the certain or sure event. The null set ϕ is the impossible event, since there are no outcomes of S in ϕ.

Two events A and B are said to be mutually exclusive if they are disjoint. That is, if $A \cap B = \phi$. In other words, A and B are mutually exclusive if they cannot occur together.

1.3.4 EXAMPLES OF EVENTS

Toss a coin three times and denote every trial of three tosses with three letters. For example, (H, T, H) represents three tosses with a head on the first toss, a tail on the second toss, and a head on the third toss.

The sample space of this random experiment is

$$S = \{(H, H, H), (H, H, T), (H, T, H), (H, T, T), (T, H, H), (T, H, T), (T, T, T), (T, T, H)\}$$

1. If the event A is "the first toss is H," it is represented by the subset $\{(H, H, H), (H, H, T), (H, T, H), (H, T, T)\}$
2. If the event B is "the third toss is H," it is represented by the subset $\{(H, H, H), (H, T, H), (T, H, H), (T, T, H)\}$
3. The event $A \cap B$ is the subset $\{(H, H, H), (H, T, H)\}$ which is the event that both A and B occur.

4. The event $A' \cup B'$ is the subset $\{(T, H, H), (T, H, T), (T, T, H), (T, T, T), (H, T, T), (H, H, T)\}$, which is the event that at most one of the events A or B occurs.

1.4 FUNCTIONS

1.4.1 DEFINITION OF A FUNCTION

A rule or a correspondence that assigns to every element of a set H a unique element of a set K is called a function f from H to K. The set H is called the domain of the function, and the set K is called the codomain of the function. The set of all elements of K that are related to the elements of H by the correspondence is called the range of the function f.

Let x represent an arbitrary, unspecified element of the set H; then it is customary to write $f(x)$ for the element of K that corresponds to x. The element $f(x)$ of K is called the value of the function f for the element x and it is read "f of x" or "f at x," or the image of x.

Set Functions and Real-Valued Functions

A function, f, whose domain is the set of real numbers or a subset of the real numbers is called a function of a real variable; and, if the range of f is also the real numbers or a subset of the real numbers, then f is called a real-valued function.
A function whose domain consists of sets of elements, such as the subsets of a sample space is called a set function.

1.4.2 IDENTICAL FUNCTIONS

Two functions, f and g, are identical if and only if they have the same domain, range, and correspondence.

Clearly, f and g are the same function if for every element $x \varepsilon H$, $f(x) = g(x)$.

1.4.3 INVERSE FUNCTIONS

Let f be a function from a sample space S to a sample space T. For a point $t \varepsilon T$, there may exist one or more than one point $x \varepsilon S$, whose image under f is t. The set of all points $x \varepsilon S$ whose image under f is t is called the inverse image of t, denoted by $f^{-1}(\{t\})$. Thus,

$$f^{-1}(\{t\}) = \{x \varepsilon S \mid f(x) = t\}$$

In general, let $B \subseteq T$. The set of all points of S for which $f(x) \varepsilon B$ is called the inverse image of B under f, denoted by $f^{1}(B)$:

$$f^{-1}(B) = \{x \varepsilon S \mid f(x) \varepsilon B\}$$

Thus,

> With every point function f, we associate a set function f^{-1}, whose domain is a class T' of subsets of T and whose range is a class of subset X of S. f^{1} is called the inverse of f. We denote
>
> $$f(X) = \{f(x) \mid x \varepsilon X, X \subseteq S\}$$
>
> $$f^{-1}(T') = \{f^{-1}(H) \mid H \subseteq T'\}$$

Evidently, $f^{-1}(T) = \{x \varepsilon S \mid f(x) \varepsilon T\} = S$

CHAPTER 2

INTRODUCTION TO PROBABILITY

2.1 PROBABILITY SPACE

For each random experiment, there is a sample space, for each sample space there are events, and for each event there is the question of the occurrence of the event.

Probability offers a measure of the occurrence of the event. Since we assign a unique real number as the probability of an event in a sample space, probability is a function which has its own distinguishing properties. The following definition lists the properties that the set of events and the probability function must satisfy.

Definition: Probability Space

A probability space is a sample space S of outcomes, a set E of subsets of S called events, and a set function P whose domain is the set E, and whose codomain is the set of real numbers such that the following two sets of axioms are satisfied :

1. Axioms for the Domain E

   B1. The sample space S is in E. That is, $S \varepsilon E$.

   B2. The empty set ϕ (the impossible event) is in E.

   B3. If a finite or countable number of events $A_1, A_2, A_3, \ldots$ belong to E, then their union belongs to E.

   B4. If an event A belongs to E, then the complement A' belongs to E.

   B5. If a finite or countable number of events $A_1, A_2, A_3, \ldots$ belong to E, then their intersection belongs to E.

2. Axioms for the Set Function P (For the codomain R of real numbers)

   PA1. For every event A of E, $0 \leq P(A) \leq 1$

PA2. $P(S) = 1$, and $P(\phi) = 0$

PA3. The probability of the union of a finite or countable number of pairwise exclusive events is the sum of the probabilities of the events; that is,

$$P(A_1 \cup A_2 \cup A_3 \cup \ldots) = P(A_1) + P(A_2) + P(A_3) + \ldots$$

$$\text{where } A_i \cap A_j = \phi \text{ for all } i \text{ and } j,\ i \neq j$$

2.2 FINITE PROBABILITY SPACES

Let $S = \{e_1, e_2, e_3, \ldots, e_n\}$ be a finite sample space, and E be the set of all subsets of S. Every event in S is a finite subset of outcomes. Since each outcome can also be considered as an elementary event, every event can be expressed as the union of elementary events. If the event $A = \{e_1, e_2\}$ then we can write $A = \{e_1\} \cup \{e_2\}$. Therefore, it is appropriate to write $e_2 \varepsilon S$ and $\{e_2\} \varepsilon E$. If we let $E_1 = \{e_1\}$, and $E_2 = \{e_2\}$, then we can write $A = \{e_1, e_2\} = E_1 \cup E_2 = \{e_1\} \cup \{e_2\}$.

The relationship between the arbitrary events $A \varepsilon E$ and the elementary events $E_1, E_2 \ldots, E_n$ along with axioms PA3, provide us with a direct way to define the probability function for the general case.

An event A which is an element of E, is an elementary event or a finite union of elementary events. Moreover, the elementary events are pairwise exclusive, therefore; the probability for A can be expressed as the sum of the probabilities for the elementary events in the union, by axiom PA3. The probability function is specified by assigning to the elementary events any set of nonnegative numbers $P_1, P_2, \ldots, P_n$ such that

$$\sum_{i=1}^{n} P_i = 1$$

That is, $P_1 + P_2 + P_3 + \ldots P_n = 1$.

If we put $P(E_i) = P_i$, then it is easily seen that

$$P(S) = \sum_{i=1}^{n} P(E_i) = \sum_{i=1}^{n} P_i = 1$$

$$0 \leq P_i \leq 1, \text{i} = 1, 2, 3, \ldots, n$$

Observe that each event A is the union of at most a finite number of elementary events E_i. Hence,

$$P(A) = \sum_{e_i \varepsilon A} P(E_i).$$

For two events A_1 and A_2 such that $A_1 \cap A_2 = \phi$ we see that

$$P(A_1 \cup A_2) = \sum_{e_i \varepsilon A_1 \cup A_2} P(E_i)$$

where e_i is an element of $A_1 \cup A_2$. Since there is no elementary event in $A_1 \cap A_2$, we have

$$P(A_1 \cup A_2) = \sum_{e_i \varepsilon A_1} P(E_i) + \sum_{e_i \varepsilon A_2} P(E_i)$$
$$= P(A_1) + P(A_2)$$

In general, $P\bigcup_{j=i}^{m} A_j = \sum_{j=1}^{m} P(A_j)$

Thus, the assignment of probabilities to the elementary events of a finite sample space is sufficient to define a probability function for the set of events of E. Observe that if there are n elementary events, then as soon as n nonnegative numbers whose sum has been selected, the probability function has been determined. For example, if $n = 2$, any two numbers p_1 and p_2 such $p_1 = 0, p_2 = 0$ and $p_1 + p_2 = 1$ are acceptable.

2.3 FINITE PROBABILITY SPACES WITH EQUALLY LIKELY OUTCOMES

Let $S = \{e_1, e_2, \ldots, e_n\}$ be a finite sample space. We wish to assign probabilities $p_1, p_2, \ldots, p_n$ to the elementary events $E_i = \{e_i\}$. That is, we wish to determine $P(E_i)$ for $i = 1, 2, \ldots, n$.

We may feel intuitively, or supported by experimentation, that each of the n outcomes, e_i, is equally likely to occur. In such cases, each of the n equally likely possible outcomes is assigned a probability of $1/n$. That is,

$$P(E_i) = 1/n, \text{ for } n = 1, 2, \ldots, n.$$

Such a finite probability space is called an equiprobable or uniform space.

Classical Approach To Probability

In a uniform probability space, the probability of an event $A \varepsilon E$ is defined as the ratio m, the number of outcomes favoring A, to n, the total number of equally likely outcomes. That is,

$$P(A) = m/n$$

This approach for assigning probabilities to events is called the classical approach or classical definition of probability.

Note that the formula $P(A) = m/n$ can be used with equiprobable space, and cannot be used in general.

2.4 OTHER APPROACHES FOR ASSIGNING PROBABILITIES TO EVENTS

Let $S = \{e_1, e_2, \ldots, e_n\}$ be a finite sample space, the events $E_i = \{e_i\}$ and E be the set of all subsets of S. Then

1. **Frequency Approach**

   In some situations, we may find it convenient to rely on the observed relative frequencies obtained from a series of repeated experiments to assign probabilities to the elementary events E_i.

   $P(E_i)$ = relative frequency of the event E_i

   The frequency approach to probability defines the probability of an event A as

$$P(A) = \text{limit } \frac{n_A}{n}$$

   where n_A is the number of times the event A occurs in n trials.

2. **Subjective Approach**

   Probabilities of the elementary events, E, may be assigned from a personal point of view. In this case, we say that the probabilities are subjective.

   It may be somewhat comforting to note that based on a large amount of data, the frequencist and the subjectivist will usually agree on the assignment of probability.

2.5 INFINITE PROBABILITY SPACES

Let $S = \{e_1, e_2, e_3, \ldots\}$ be a countably infinite sample space, $E_i = \{e_i\}$ where $i = 1, 2, 3, \ldots, \infty$, and E be the set of all subsets of S. As in the finite case, we can obtain a probability space by assigning to each outcome $e_i \varepsilon S$ a real number p_i, called its probability. That is $P(E_i) = p_i$, for $i = 1, 2, 3, \ldots, \infty$ such that

$$0 \leq p_i \leq 1, \text{ and } p_1 + p_2 + p_3 + \ldots = \sum_{i=1}^{\infty} p_i = 1$$

The probability $P(A)$ of any event $A \varepsilon E$ is then given by the sum of the probabilities of its points.

$$P(A) = \sum_{e_i \varepsilon A} P(E_i)$$

For example, a coin is tossed until the first time that head turns up. Let x be the first time a head turns up. Then

$$S = \{1, 2, 3, \ldots\}.$$

The probability that a head comes up on the first toss is 1/2 and the probability that a head comes up the first time on the second toss is $1/2^2$ which is equal to 1/4 and so forth. That is

$$P(n) = 1/2^n, \text{ for } n = 1, 2, 3, \ldots$$

$$\text{and} \quad P(x) = 1/2 + 1/4 + 1/8 + \ldots = 1$$

> **Definition**: Discrete and Nondiscrete Probability Spaces
> A finite or countably infinite probability space is said to be discrete, and an uncountable space is said to be nondiscrete.

2.6 PROPERTIES OF THE PROBABILITY FUNCTION

Let S be a sample space, E be the set of all subsets of S, and $A, B, C, A_1, A_2, A_3, \ldots A_n$ denote arbitrary events of the set E.

Then, using the axioms stated in section 2.1 for the probability function P, we may state the following rules which often simplify the determination of probabilities:

1. The probability of the impossible event is zero. That is, $P(\phi) = 0$.
2. $P(A) \leq 1$ for all $A \varepsilon E$.
3. The probability of the difference of two events is given by $P(B - A) = P(B) - P(A \cap B)$.
4. The probability of the complement of the event A is $1 - P(A)$. That is, $P(A') = 1 - P(A)$.
5. The probability of the union of two events A and B is given by $P(A \cup B) = P(A) + P(B) - P(A \cap B)$.
6. If event A is a subevent of event B, then the probability of event A is less than the probability of event B; that is, if $A \subseteq B$, then $P(A) \leq P(B)$.
7. The probability of the union of any three elements A, B, and C in S is given by $P(A \cup B \cup C) = P(A) + P(B) + P(C) - P(A \cap B) - P(A \cap C) - P(B \cap C) + P(A \cap B \cap C)$.

8. Addition Law For Mutually Exclusive Events: If $A_1, A_2, \ldots A_n$, are mutually exclusive events in S, then

$$P(A_1 \cup A_2 \cup \ldots \cup A_n) = \sum_{i=1}^{n} P(A_i).$$

CHAPTER 3

COUNTING METHODS

3.1 SAMPLING AND COUNTING

There are many instances in the application of probability theory where it is desirable and necessary to count the outcomes in the sample space and the outcomes in an event. For example, in the special instance of a uniform probability function, the probability of an event is known when the number of outcomes that comprises the event is known; that is, as soon as the number of outcomes in the subset that defines the event is known.

> If a sample space $S = \{e_1, e_2, \ldots, e_n\}$ contains n simple events, $E_i = \{e_i\}, i = 1, 2, \ldots, n$, then using a uniform probability model, we assign probability $1/n$ for each point in S; that is, $P(E_i) = 1/n$. To determine the probability of an event A, we need,
>
> 1. The number of possible outcomes in S.
> 2. The number of outcomes in the event A.
>
> Then,
>
> $$P(A) = \frac{\text{number of outcomes corresponding to } A}{\text{number of possible outcomes in } S}$$

Frequently, it may be possible to enumerate fully all the sample space points in S and then count how many of these correspond to the event A. For example, if a class consists of just three students, and the instructor always calls on each student once and only once during each class, then if we label the students 1, 2, and 3, we can easily enumerate the points in S as

$$S = \{(1, 2, 3), (1, 3, 2), (2, 1, 3), (2, 3, 1), (3, 1, 2), (3, 2, 1)\}$$

Assume that the instructor chooses a student at random, it would seem reasonable to adopt a uniform probability model and assign probability 1/6 to each point in S. If A is the event that John is selected last, then

$$\text{John} = B$$

$$A = \{(1, 2, 3), (2, 1, 3)\}$$

$$P(A) = 2/6 = 1/3$$

It would be most unusual for a class to consist of only three students. The total enumeration of the sample space becomes more complicated even if we increase the class size to 6 students. To deal with these situations in which the sample space contains a large number of points, we need to have an understanding of basic counting or combinatorial procedures.

3.2 THE FUNDAMENTAL PRINCIPLE OF COUNTING

Suppose a man has four ways to travel from New York to Chicago, three ways to travel from Chicago to Denver, and six ways to travel from Denver to San Francisco, in how many ways can he go from New York to San Francisco via Chicago and Denver?

If we let A_1 be the event "going from New York to Chicago," A_2 be the event "going from Chicago to Denver," and A_3 be the event "going from Denver to San Francisco," then because there are four ways to accomplish A_1, 3 ways to accomplish A_2, and 6 ways to accomplish A_3, the number of routes the man can follow is

$$(4) \times (3) \times (6) = 72$$

We can now generalize these results and state them formally as the fundamental principle or multiplication rule of counting.

Fundamental Principle of Counting

If an operation consists of a sequence of k separate steps of which the first can be performed in n_1 ways, followed by the second in n_2 ways, and so on until the kth can be performed in n_k ways, then the operation consisting of k steps can be performed in

$$n_1 \times n_2 \times n_3 \cdot \ldots \cdot n_k$$

ways.

3.2.1 TREE DIAGRAMS

> A tree diagram is a device that can be used to list all possible outcomes of a sequence of experiments where each experiment can occur only in a finite number of ways.

The following tree diagram lists the different ways three different flavors of ice cream, chocolate (c), vanilla (v), and strawberry (s), can be arranged on a cone, with no flavor used more than once.

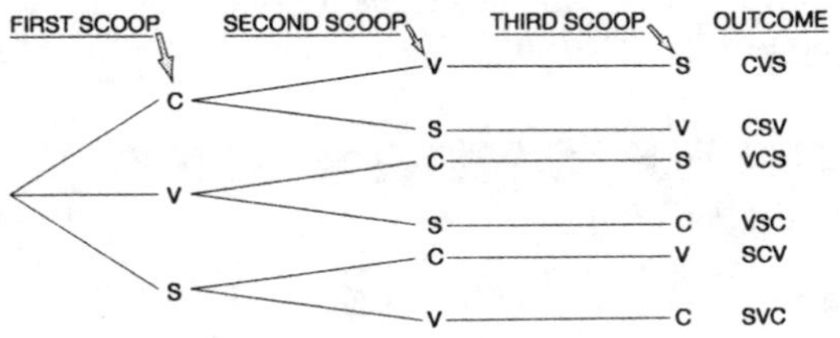

The tree starts with three branches in the first stage, representing the three possibilities for first stage. For each outcome at the first stage, there are two possibilities at the second stage. Then, for each outcome in the second stage, there is only one possibility at the third stage. Consequently, there are $3 \times 2 \times 1$, or 6, different arrangements.

Using a tree diagram, we can develop the sample space for an experiment consisting of tossing a fair coin and then rolling a die as follows:

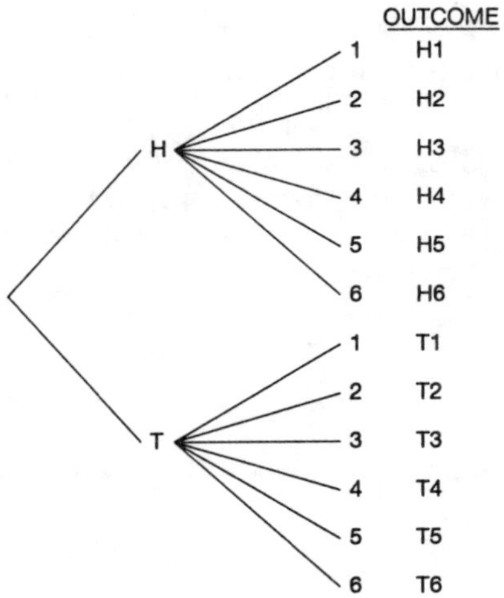

3.3 FACTORIAL NOTATION

Consider how many ways the owner of an ice cream parlor can display ten ice cream flavors in a row along the front of the display case. The first position can be filled in ten ways, the second position in 9 ways, and the third position in 8 ways, and so on. By the fundamental counting principle, there are

$$(10) \times (9) \times (8) \times (7) \times ... \times (2) \times (1)$$

or 3,628,800 ways to display the flavor. If there are 16 flavors, there would be $(16) \times (15) \times (14) \times ... \times (3) \times (2) \times (1)$ ways to arrange them. In general

> If n is a natural number, then the product from 1 to n inclusive is denoted by the symbol $n!$ (read as "n factorial" or as "factorial n" and is defined as
>
> $$n! = n(n - 1)(n - 2) \ldots (3)(2)(1)$$
>
> where n is a positive natural number.

There are two fundamental properties of factorials:

> 1. By definition, $0! = 1$
> 2. $n(n - 1)! = n!$

For example, $(6)(5!) = 6!$

3.4 COUNTING PROCEDURES INVOLVING ORDER RESTRICTIONS (PERMUTATIONS)

Suppose a class consists of 5 students. The instructor calls on exactly three students out of the 5 students during each class period to answer three different questions. To apply the uniform probability model, we need to know how many points there are in the sample space S. Note that each point in S is an ordered triplet; that is, the point (3, 5, 1) is different from the point (5, 3, 1). The same three people were called upon to respond, but the order of response is different. Such arrangement is referred to as a permutation.

> A permutation of a number of objects is any arrangement of these objects in a definite order.

For example, if a class consists of 3 students, then there are $3 \times 2 \times 1 = 6$ ways in which the students might be called upon.

In general, if the class had consisted of n students and all of them had been called upon, then the responses could have taken place in

$$n(n - 1)(n - 2) \ldots (3)(2)(1)$$

ways. Hence,

> The number of permutations of a set of n distinct objects, taken all together, is $n!$

In our example of the class consisting of five students, only three students were to be called on to respond; that is, we are interested in an ordered subset.

> An arrangement of r distinct objects taken from a set of n distinct objects, $r \leq n$, is called a permutation of n objects taken r at a time. The total number of such orderings is denoted by nPr, and defined as
>
> $$nPr = \frac{n!}{(n-r)!}$$

In our example, $n = 5$, $r = 3$,

$$5P3 = \frac{5!}{(5-3)!} = 5 \times 4 \times 3 = 60$$

If we have n items with r objects alike, then the number of distinct permutations taking all n at a time is

$$\frac{n!}{r!}$$

In general,

> In a set of n elements having r_1 elements of one type, r_2 elements of a second type and so on to r_k elements of a kth type, then the number of distinct permutations of the n elements, taken all together, is given by
>
> $$nPn = \frac{n!}{r_1!r_2!r_3!\ldots r_k!}$$
>
> $$\text{where } \sum_{i=1}^{k} r_i = n$$

For example, the number of ways a group of 10 of which 6 are females and 4 are males can line up for theatre tickets, if we are interested only in distinguishing between sexes, is given by

$$10P10 = \frac{10!}{6!\,4!} = 210$$

3.5 COUNTING PROCEDURES NOT INVOLVING ORDER RESTRICTIONS (COMBINATIONS)

Suppose that a class of 12 students selects a committee of 3 to plan a party. A possible committee is John, Sally, and Joe. In this situation, the order of the three is not important, because the committee of John, Sally, and Joe, is the same as the committee of Sally, Joe, and John.

When choosing committee members and in other cases of selection where order is not important, we are interested in combinations, not permutations. In general,

> Combination
>
> A subset of r objects selected without regard to order from a set of n different objects, $r \leq n$, is called a combination of n objects taken r at a time. The total number of combinations of n things taken r at a time is denoted by nCr or $\binom{n}{r}$ and is defined as
>
> $$nCr = \binom{n}{r} = \frac{n!}{r!(n-r)!}$$

In our example, the number of possible committees that could plan the party can be calculated by

$$12C3 = \binom{12}{3} = \frac{12!}{3!(12-3)!} = 220$$

3.6 THE BINOMIAL THEOREM AND COEFFICIENTS

The combination symbols we have discussed in the previous section are extremely useful in connection with the expansion of powers of binomial and multinomial terms.

> The Binomial Theorem
>
> If n is any positive integer, then
>
> $$(a+b)^n = \binom{n}{0}a^n b^0 + \binom{n}{1} + a^{n-1}b^1 + \binom{n}{2}a^{n-2}b^2 + \ldots + \binom{n}{r}a^{n-r}b^r +$$
>
> $$+ \ldots + \binom{n}{n}9^0 b^n$$

$$= \sum_{r=0}^{n} \binom{n}{r} a^{n-r} b^r$$

For example,

$$(a+b)^2 = \binom{2}{0} a^2 + \binom{2}{1} ab + \binom{2}{2} b^2$$

$$= a^{2+} 2ab + b^2$$

$$(a+b)^3 = \binom{3}{0} a^3 + \binom{3}{1} a^2 b + \binom{3}{2} ab^2 + \binom{3}{3} b^3$$

$$= a^3 + 3a^2 b + 3a\, b^2 + b^3$$

Binomial Coefficients

The coefficients in the algebraic expansion of the binomial expression $(a + b)^n$ are the integers

$$= \frac{n!}{r!(n-r)!},$$

where r and n are positive integers with $r \leq n$, are called binomial coefficients.

For example, in the expansion of $(a + b)^3$, the binomial coefficients are: $\binom{3}{0}$, $\binom{3}{1}$, $\binom{3}{2}$, and $\binom{3}{3}$, or 1, 3, 3, and 1.

3.6.1 PROPERTIES OF BINOMIAL EXPANSIONS

The expansions in positive integral powers, n, of $(a + b)^n$ have the following properties:

1. There are $(n + 1)$ terms in each expansion.
2. The sum of the exponents for a and b in each term is n.
3. The exponents of a decrease from n to 0 while the exponents of b increase from 0 to n.
4. Since $\binom{n}{n-r} = \binom{n}{r}$, or in other words if $r + k = n$, then $\binom{n}{r} = \binom{n}{k}$, it follows that the coefficient of the rth term is $\binom{n}{r}$, where r is

the exponent of either a or b.

5. Coefficients of terms equidistant from the ends are the same.

3.6.2 PASCAL'S RULE (TRIANGLE)

$$\binom{n+1}{r} = \binom{n}{r-1} + \binom{n}{r}, \text{ for } 1 \le r \le n$$

Pascal's triangle is useful for constructing tables of binomial coefficients. If, for example, we know the coefficients for $n = 10$, we can determine those for $n = 11$ as follows :

$$\binom{11}{r} = \binom{11}{r-1} = \binom{10}{r}$$

Pascal's triangle has the following properties:

1. The first number and the last number in all rows is 1.
2. Every other number in the array is obtained by adding the two numbers appearing directly above it.

For example, the coefficient of successive powers of $(a + b)$ can be arranged in Pascal's triangular array of numbers as follows:

$(a + b)^0$	1
$(a + b)^1 = a + b$	1 1
$(a + b)^2 = a^2 + 2ab + b^2$	1 2 1
$(a + b)^3 = a^3 + 3a^2b + 3ab^2 + b^3$	1 3 3 1
$(a + b)^4 = a^4 + 4a^3b = 6a^2b^2 + 4ab^3 + b^4$	1 4 6 4 1
....	....
....	....

3.7 REPEATED BERNOULLI TRIALS

Let A and B be two events in a sample space S. We say that A and B are independent if and only if the occurrence of one of the events does not affect the probability of the occurrence of the other.

If a random experiment being repeated has only two outcomes, such as; {male, female}, or {on, off}, or {head, tail}, then we have a particularly important case of repeated trials, known as Bernoulli trials. Intuitively, a sequence of Bernoulli trials is no more complicated than 100 tosses of a coin. If we label the two outcomes as "success" (s), or "failure (f) with

$P\{(s)\} = p$, and $P\{(f)\} = 1 - p = q$, then in a binomial experiment (an experiment with only two outcomes), we are interested in the probability of exactly x successes in n Bernoulli trials.

> Bernoulli Trials
>
> Repeated trials of a random experiment are called Bernoulli trials if
>
> a. There are only two outcomes possible for each trial.
>
> b. The probabilities of the outcomes are the same for each trial.
>
> c. The trials are independent.

Note that each point in the sample space consisting of x s's and $(n - x)$ f's has the probability

$$p^x q^{(n-x)}$$

3.8 MULTINOMIAL EXPANSIONS

The Multinomial Theorem

Let $n_1, n_2, n_3, \ldots, n_k$ and n be nonnegative integers such that $n_1 + n_2 + n_3 + \ldots + n_k = n$. Then,

$$(a_1 + a_2 + a_3 + \ldots + a_k)^n = \sum_{n_1=0}^{n} {}^{n} \sum_{n_2=0}^{n-n_1} \sum_{n_3=0}^{n-n_1-n_2} \sum_{n_k=0}^{n-n,n_2-\ldots n_{k-2}} (n_1, n_2, \ldots, n - n_1 - \ldots - n_{k-1}) \left(a_1^{n_1} a_2^{n_2} a_3^{n_3} \ldots a_k^{n-n_1-\ldots n_{k-1}}\right)$$

$$= \sum_{n_1+n_2\ldots+n_k=n} \left(n_1, n_2^n, \ldots, n_k\right) a_1^{n_1} \times a_2^{n_2} \times a_3^{n_3} \ldots a_k^{n_k}$$

where $a_1, a_2, \ldots, a_n$ are real numbers

The expression $(n_1, n_2, \ldots, n_k)$ is defined as follows:

$$\binom{n}{n_1, n_2, n_3 \ldots, n_k} = \frac{n!}{n_1! n_2! n_3! \ldots n_k!}$$

Multinomial Coefficients

The numbers $\frac{n!}{n_1!n_2!n_3!\ldots n_k!}$ are called the multinomial co-efficients in view of the multinomial theorem.

For example $\binom{10}{3,\ 4,\ 2,\ 1} = \frac{10!}{3!\ 4!\ 2!\ 1!} = 12,600$

Note that $\frac{n!}{n_1!n_2!\ldots n_k!} = \binom{n}{n_1}\binom{n-n_1}{n_2}\binom{n-n_1-n_2}{n_3}\ldots\binom{n-n_1-n_2-\ldots-n_{k-2}}{n_{k-1}}$

3.9 PROBABILITY PROBLEMS

The counting methods can be used to determine probabilities of events that arise from different types of probability problems. The procedure for solving problems usually involves four steps:

1. Select a sample space for the random experiment.
2. Assign a probability function, *P*, to the events of the sample space.
3. Define the events in question as a subset of the sample space.
4. Determine the value of the probability function for this event.

In the following examples, let $N(S)$ denote the number of elements in the sample space *S*.

Example (1)

Winning tickets for first, second, and third prizes are drawn from a bowl containing 450 tickets numbered from 1 to 450 inclusive. Find the probability that ticket 1 won a prize.

The 450 tickets is a set from which 3 tickets are to be selected; with the 3-tuple of numbers the prize winners. Thus, the sample space consists of permutation of 450 things taken 3 things at a time outcome; that is $N(S) = 450P3$. The event *A* that ticket 1 wins a prize is the subset of outcomes with a 1 in one of the three spots. If 1 is the first component, there are $449P2$ choices for the second and third components. If 1 wins second prize, there are $449P2$ choices for first and third prizes. Since the same reasoning applies if ticket 1 wins third prize, there are $(3)(449P2)$ outcomes in event *A*. So $N(A) = (3)(449P2)$. Hence,

$$P(A) = \frac{N(A)}{N(S)} = \frac{(3)(459P2)}{450P3} = \frac{3.449!.447!}{447!\ 540!} = \frac{3}{450} = \frac{1}{150}$$

Example (2)

A committee of 5 is to be selected at random from a group of 10 men and 8 women. What is the probability that there will be exactly 3 men on the committee?

The 10 men and 8 women constitute a set of 18 elements with 10 elements of one kind and 8 elements of another kind. So, the sample space S consists of $\binom{18}{5}$ outcomes; that is $N(S) = \binom{18}{5}$. The event A is the subset of S with 3 men and 2 women in the selection. Now this selection can be broken into two parts with the selection of the men one task and the selection of the women a second task. The 3 men represent a combination of 10 objects taken 3 at a time; that is, $\binom{10}{3}$. The 2 women represent a combination of 8 objects taken 2 at a time; that is, $\binom{8}{2}$. Thus there are $\binom{10}{3} \times \binom{8}{2}$ elements in event A and

$$P(A) = \frac{\binom{10}{3}\binom{8}{2}}{\binom{18}{5}} = \frac{20}{51}$$

3.10 RANDOM SAMPLING

Essentially, random sampling is an application of the uniform probability model. Using the concepts of the uniform model, we define a random sample as follows.

> Random Sample
>
> A sample is said to be a random sample if all possible samples, of a particular size, chosen under some specified selection schemes, have the same probability of being chosen.

The following property is referred to as the "equivalence law of ordered sampling."

> Equivalence Law of Ordered Sampling
>
> If a random sample of size k is drawn from a population of size N, then on any particular one of the k draws each of the N items has the same probability $1/N$ of being selected. The probability that a specified item is included in a sample of size k taken from a population of size N is k/N.

CHAPTER 4

CONDITIONAL PROBABILITY AND INDEPENDENCE

4.1 CONDITIONAL PROBABILITY

Frequently we are interested in probabilities concerning part, rather than all, of a sample space. For example, the probability that a person chosen at random from a population has blue eyes is different from the probability of blue eyes in a subpopulation of people with blonde hair. Probabilities associated with these subpopulations are called conditional probabilities.

> Conditional Probability For Two Events
>
> Let A and B be two events of a probability space S with probability function P, and $P(B) > 0$, then the conditional probability of event A, given that event B has occurred, is denoted by $P(A/B)$ and is defined
>
> $$P(A/B) = \frac{P(A \cap B)}{P(B)}$$
>
> for $P(B) \neq 0$. If $P(B) = 0$, then $P(A/B)$ is not defined.

The conditional probability for two events in a sample space S is defined in terms of the probability function P for S. Hence, all axioms and theorems for general probability functions hold for the conditional probability function. The following is a list of such axioms and theorems. In those statements, S denotes a sample space for which $P(S) = 1$, and A, B, E and F are arbitrary events in S with $P(F) > 0$.

> Properties of the Conditional Probability Function
>
> 1. $0 \leq P(E/F) \leq 1$
> 2. $P(S/F) = 1$
> 3. $P[(A \cup B)/F] = P(A/F) + P(B/F)$, if $A \cap B = \phi$.
> 4. $P[(A \cup B)/F] = P(A/F) + P(B/F) - P[(A \cap B)/F]$
> 5. $P(E'/F) = 1 - P(E/F)$

4.2 CONDITIONAL PROBABILITY APPLICATIONS

1. Let A and B be two nonempty events in a sample space S with $0 < P(A) < 1$. Then

$$P(B) = P(B/A) \times P(A) + P(B/A') \times P(A')$$

For example, two urns, urn 1 contains two black balls and three white balls; and urn 2 contains two black balls and one white ball. An urn is chosen at random and a ball picked from this urn. If we let A be the event "selection of urn 1," then A' is the event "selection of urn 2." Also, $P(A) = P(A') = 1/2$. If B is the event "picking a white ball," we obtain

$$P(B/A) = 3/5, \text{ and } P(B/A') = 1/3$$

Consequently, we have $P(B) = P(B/A) \times P(A) + P(B/A') \times P(A')$

$$= (3/5)(1/2) + (1/3)(1/2) = 7/15$$

2. If $B_1, B_2, \ldots, B_n$ is a partition of a sample space S, where $P(B_i) > 0$, $i = 1, 2, \ldots, n$, and if A is any event in S, then

$$P(A) = \sum_{i=1}^{n} P(A/B_i) \times P(B_i)$$

For example, in a small city there are four voting subdivisions, identified as ward 1, ward 2, ward 3, and ward 4. The political composition of the wards with regard to the proportion showing preference for the Democratic Party is as follows:

ward 1: 0.45, ward 2: 0.35, ward 3: 0.45, ward 4: 0.55

A voter is selected from the total voting population by first selecting a ward, then choosing a voter, both selections being random . What is the probability that a voter with Democratic preference is selected?

The sample space is the set of all voters in the city. Let B_i be the event "a voter lives in ward i," $i = 1, 2, 3, 4$, and let A be the event "a voter is a Democrat." We want to find $P(A)$. Now, $P(B_i) = 1/4$, $i = 1, 2, 3, 4$, and

$$P(A/B_1) = 0.45, \quad P(A/B_2) = 0.35, \quad P(A/B_3) = 0.45,$$

and $P(A/B_4) = 0.55$

Since a voter does not legally live in 2 wards, we can assume that all voters in the city belong to the set S given

$$S = B_1 \cup B_2 \cup B_3 \cup B_4$$

where $B_i \cap B_j = \phi$ for $i \times j = 1, 2, 3, 4$, and $i \neq j$. Thus, $\{B_1, B_2, B_3, B_4\}$ is a partition of S. Hence, we may write

$$P(A)=\sum_{j=1}^{4} P\left(A/B_j\right)\times\left(B_j\right)=(.45+.35+.45+.55)(1/4)=0.45$$

4.3 MULTIPLICATION RULE AND ASSIGNING PROBABILITIES

Let A and B be two events of a sample space S with $P(B) \neq 0$. From the definition of conditional probability we know that

$$P(A/B)=\frac{P(A\cap B)}{P(B)}$$

> Multiplication Rule of Probability
>
> If we multiply both sides of the equation
>
> $$P(A/B)=\frac{P(A\cap B)}{P(B)}$$
>
> by $P(B)$, we obtain
>
> *P(A « B) = P(A/B) ¥ P(B)*
>
> This is called the multiplication rule or multiplication law of probability.

The multiplication rule can be easily extended for any number of events.

> Generalized Multiplication Rule
>
> For any events $A_1, A_2, \ldots, A_n$,
>
> $$P(A_1 \cap A_2 \cap \ldots \cap A_n) = P(A_1) \times P(A_2/A_1) \times P[A_3/(A_1 \cap A_2)] \ldots .P[A_n/(A_1 \cap A_2 \cap A_3 \cap \ldots \cap A_{n-1})]$$
>
> If $n = 3$, then we obtain,
>
> $$P(A_1 \cap A_2 \cap A_3) = P(A_1) \times P(A_2/A_1) \times P(A_3/A_1 \cap A_2)$$

4.4 STAGEWISE EXPERIMENTS

Many experiments can be decomposed into stages or trials in which each stage or trial has a finite number of outcomes with given probabilities. A convenient way of describing such a process and computing the probability of any event in the resulting sample space is a tree diagram, which gives a pictorial view of the stages. Then the multiplication rule can be used to compute the probability that the result represented by any given path of the tree does occur. That is, to find the probability of an intersection, we multiply the corresponding branch (path) probabilities.

For example, a third year Arts student finds at the end of his final year that he is lacking a science course needed for graduation. He examines the summer school offerings and finds that he can take a course in mathematics (m), chemistry (c), or computer science (cs). On the basis of his interest, he assigns a probability of 0.1, 0.6, and 0.3 to the events of choosing each of these subjects. After considering his past performance, his advisor estimates the probability of his passing (p) the mathematics course as 0.8, the chemistry course as 0.7, and the computer science course as 0.75.

a) What is the probability of his passing?

b) If at the end of the summer, you hear that he has graduated, what is the probability that he took

   i) the mathematics course?

   ii) the chemistry course?

   iii) the computer science course?

The following tree diagram summarizes the given data.

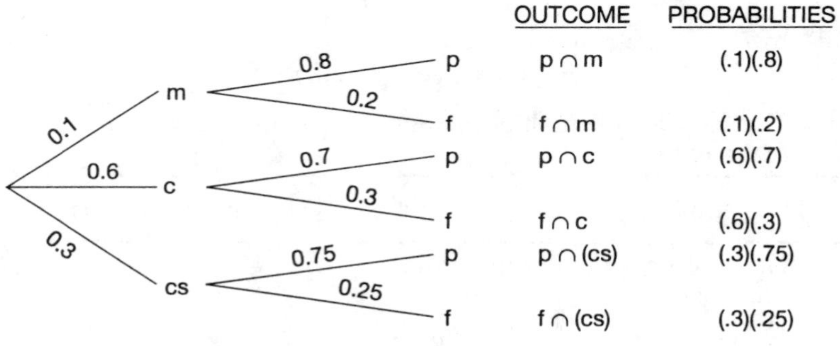

a) $P(p) = P(p \cap m) + P(p \cap c) + P(p \cap (cs))$

$= (.1)(.8) + (.6)(.7) + (.3)(.75) = 0.725$

b) Using the definition of conditional probability, we obtain

i) $P(m|p) = \dfrac{(0.1)(0.8)}{0.725} = 0.11$

ii) $P(c|p) = \dfrac{0.6(0.7)}{0.725} = 0.58$

iii) $p((cs)/p) = \dfrac{0.3(0.75)}{0.725} = 0.31$

4.5 INDEPENDENT EVENTS

4.5.1 TWO INDEPENDENT EVENTS

Let A and B be any 2 events of the same sample space S. Then events A and B are said to be independent if

$$P(A \cap B) = P(A) \times P(B).$$

Two events that are not independent are called dependent.

Sometimes independent events are referred to as "statistically independent" or "stochastically independent."

Remark

If A and B are independent events, then

a. A' and B'

b. A and B'

c. A' and B

are independent events.

4.5.2 INDEPENDENCE OF MORE THAN TWO EVENTS

Mutually Independent Events

Let $A_1, A_2, \ldots, A_n$ be n events defined on the same space S. The n events $A_1, A_2, \ldots, A_n$ are said to be mutually independent if the joint probability of every combination of the events taken any number, r, at a time, $r \leq n$, is equal to the product of their individual probabilities. That is, for any subcollection A_{i1}, $A_{i2}, \ldots, A_{ik}$, $2 \leq k \leq n$,

$$P\left(\bigcap_{j=1}^{k} A_{ij}\right) = \pi_{j-1}^{k} P\left(A_{ij}\right),$$

where π stands for the product.

If $n = 3$, mutual independence means

$$P(A_1 \cap A_2 \cap A_3) = P(A_1) \times P(A_2) \times P(A_3)$$

$$P(A_1 \cap A_2) = P(A_1) \times P(A_2)$$

$$P(A_1 \cap A_3) = P(A_1) \times P(A_3)$$

$$P(A_2 \cap A_3) = P(A_2) \times P(A_3)$$

All four of these conditions must hold. In the case in which only the last three conditions hold, the events A_1, A_2, and A_3 are said to be "pairwise independent". Note that pairwise independence does not imply mutual independence. For example, toss a penny and a nickel and define the events: A_1 "head on penny," A_2 "head on nickel," and A_3 "coins match." Then

$$P(A_1) = P(A_2) = P(A_3) = 1/2$$

and

$$P(A_1 \cap A_2) = P(A_1 \cap A_3) = P(A_2 \cap A_3) = 1/4$$

Thus, A_1, A_2, and A_3 are pairwise independent. However,

$P(A_1 \cap A_2 \cap A_3) = 1/4 \neq P(A_1) \times P(A_2) \times P(A_3)$

Hence, although the events are pairwise independent, they are not mutually independent.

4.6 PROBABILITIES ASSOCIATED WITH MUTUALLY INDEPENDENT EVENTS

Under the assumption of independence, the multiplication law for two independent events can be generalized for n independent events as follows.

> Multiplication Law For n Independent Events
>
> Let $A_1, A_2, \ldots, A_n$ be events defined on the same sample space S.
>
> If $A_1, A_2, \ldots, A_n$ are independent, then
>
> $P(A_1 \cap A_2 \cap A_3 \cap \ldots \cap A_n) = P(A_1) \times P(A_2) \times P(A_3) \ldots P(A_n)$

For example, if a box contains r red marbles and b black marbles, and a game consists of drawing with replacement until a red marble appears, what is the probability of winning on the kth draw?

Since the draws are with replacement, we can visualize the experiment as consisting of k independent trials. For each trial,

$$P(red) = \frac{r}{r+b}, \text{ and } P(black) = \frac{b}{r+b}$$

winning on the kth draw results from the first $(k - 1)$ draws being black marbles and the kth being a red marble. Hence,

$$P(\text{winning on kth draw}) = \left(\frac{r}{r+b}\right)^{k-1}\left(\frac{r}{r+b}\right)$$

4.7 INDEPENDENT OR REPEATED TRIALS

A fair die with three faces marked with a 3, two faces marked with a 2 and one face marked with a 1 is tossed. A sample space is the set $S = \{1, 2, 3\}$, and if the elementary event E_i corresponds to the outcome i for $i = 1, 2, 3$, a probability function for S is $P(E_1) = 1/6$, $P(E_2) = 1/3$, and $P(E_3) = 1/2$. If we toss the die twice, and assume independence for the tosses, then the new sample space T will be the set $T = \{(1,1), (1,2), (1,3), (2,1), (2,2), (2,3), (3,1), (3,2), (3,3)\}$. For the elementary event $\{(2, 3)\}$ of T,

$$P\{(2,3)\} = (1/3)(1/2) = 1/6$$

In general, we can define the probability space of two repeated independent trials as follows.

> Probability Space Of Two Repeated Independent Trials
>
> For a probability space with a countable sample space $S = \{e_1, e_2, e_3, \ldots\}$, elementary events $E_1, E_2, E_3, \ldots$, and probability function $P(E_i) = p_i$ for $i = 1, 2, \ldots$, the probability space of two repeated independent trials consists of a sample space, T, of 2-tuples (e_i, e_j) to which the probabilities
>
> $$P(\{e_i, e_j\})\ p_i p_j$$
>
> are assigned.

If A_1 and A_2 are two events of a probability space, T, of two repeated trials such that A_1 depends only on the first trial and A_2 depends only on the second, then the events A_1 and A_2 are independent.

4.7.1 PROBABILITY SPACES OF *n* REPEATED INDEPENDENT TRIALS

> For a probability space of a countable sample space $S = \{e_1, e_2, \ldots\}$, elementary events $E_1, E_2, E_3, \ldots, (E_i = e_i)$, and probability function $P(E_i) = p_i$, for $i = 1, 2, 3, \ldots$, the probability space, T, of n repeated independent trials consists of a sample space with n-tuples $(e_1, e_2, e_3, \ldots e_n)$ to which the probabilities
>
> $$P(e_1, e_2, \ldots e_n) = p_1 p_2 p_3 \cdots p_n$$
>
> are assigned.

4.7.2 PROBABILITY SPACES OF DIFFERENT EXPERIMENTS

A succession of different experiments can be treated in the same way as repeated trials of one experiment. The essential feature is the assumption of independence.

For $S_1 = \{e_1, e_2, e_3, \ldots, e_n\}$ and $S_2 = \{f_1, f_2, \ldots, f_n\}$ are two sample spaces; $E_1, E_2, \ldots, E_n$ $(E_i = \{e_i\})$, $F_1, F_2, \ldots, F_n$ $(F_i = \{f_i\})$ are elementary events. Then the probability space, T, of pairs of events that result from executing experiments 1 and 2, and with probabilities assigned to the pairs by

$$P(\{e_i, F_i\}) = P(E_i) \times P(F_i)$$

represents the compound experiment of two successive independent experiments.

For example, if a fair coin is tossed and a ball is selected from an urn with 7 red, 4 white, and 3 green balls, and if we assume that the experiments are independent, then the probability of an even number on the die and a white ball selected is the product (1/2)(4/14) = 1/7.

Generalizing this notion to n different experiments can be done exactly as we generalized from two repetitions to n repeated experiments.

4.8 BERNOULLI TRIALS AND THE BINOMIAL AND MULTINOMIAL PROBABILITY FUNCTIONS

4.8.1 BINOMIAL PROBABILITY FUNCTION

If a random experiment consists of n Bernoulli trials with p the probability of success on each trial, and l-p the probability of failure on each trial, then the probability function for the sample space of the number of success x in n trials is given by

$$P(x) = \binom{n}{x} p^x q^{n-x}$$

where $q = 1 - p$, and $x = 0, 1, 2, 3, \ldots, n$. Note here that the probabilities are the terms of the expansion $(p + q)^n$.

A sequence of repeated trials with more than two outcomes is often converted to the case of Bernoulli trials by considering the occurrence of event A as a success and the occurrence of A' as failure, with $p = P(A)$ and $q = P(A')$. For example, in measuring bolts produced by a machine, a bolt is considered defective if the length does not fall between two certain limits. Thus, if we are counting the number of defective bolts, a bolt is considered a "success" if it is defective. The bolts are thereby separated into two classes, A = {defective bolts} and A' = {good bolts}.

4.8.2 MULTINOMIAL PROBABILITY FUNCTION

Consider the generalized Bernoulli trials. Here we have a sequence of independent trials, and on each trial the result is exactly one of the k possibilities $b_1, b_2, \ldots, b_k$. On a given trial, let b_i occur with probability p_i, $i = 1, 2, \ldots, k$, $p_i \geq 0$, and $\sum_{i=1}^{k} p_i = 1$.

The sample space S = all k^n ordered sequences of length n with components $b_1, b_2, \ldots, b_k$. For example, if $x = (b_1, b_3 b_2, b_2 \ldots)$, then b_1 occurs on trial 1, b_3 occurs on trial 2, b_2 occurs on trials 3 and 4, and so on. If we assign to the point

$$x = \underbrace{b_1 b_1 \ldots b_1}_{n_1} \underbrace{b_2 b_2 \ldots b_2}_{n_2} \underbrace{b_k b_k \ldots b_k}_{n_k}$$

the probability $p_1^{n_1} p_2^{n_2} \ldots p_3^{n_k}$, this is the probability assigned to any sequence having n_i occurrences of b_i, $i = 1, 2, \ldots, k$, then the number of sequences having exactly n_1 occurrences of b_1, n_2 occurrences of $b_2, \ldots$ n_k occurrences of b_k is given by

$$\frac{n!}{n_1!n_2!\,n_k!}$$

Multinomial Probability Function

The probability that b_1 will occur n_1 times, b_2 will occur n_2 times, … , and b_k will occur n_k times is given by

$$P(n_1, n_2, \ldots, n_k) = \binom{n}{n_1, n_2, \ldots n_k} p_1^{n_1} p_2^{n_2} \cdots p_k^{n_k}$$

$$= \frac{n!}{n!n!n!\ldots n!} \times p_1^{n_1} p_2^{n_2} \cdots p_k^{n_k}$$

The function $P(n_1, n_2, \ldots, n_k)$, where $n_1, n_2, \ldots, n_k$ are nonnegative integers such that $n_1 + n_2 + \ldots + n_k = n$ is called the multinomial probability function.

Note that when $k = 2$, the generalized Bernoulli trials reduce to ordinary Bernoulli trials where b_1 = "success," b_2 = "failure," $p_1 = p$, $p_2 = q$, $n_1 = k$, $n_2 = n - k$, and

$$\frac{n!}{n_1!n_2!} = p_1^{n_1} p_2^{n_2} = \binom{n}{k} p^n q^{n-k}$$

= probability of k successes in n trials.

4.9 POSTERIOR PROBABILITIES: BAYE'S RULE

An application of conditional probability is the famous formula known as Baye's theorem.

Baye's Theorem

Let $B_1, B_2, \ldots, B_n$ be a partition of a sample space S. If the unconditional probabilities $P(B_i)$, $i = 1, 2, \ldots, n$, and the conditional probabilities $P(A/B_i)$, $i = 1, 2, \ldots, n$, are known, then

$$P(B_i/A) = \frac{P(A/B_i) \times P(B_i)}{\sum_{i=1}^{n} P(A/B_i) \times P(B_i)}$$

For example, suppose that we have 3 urns, urn 1 has 3 red and 4 blue chips; urn 2 has 4 red and 3 blue chips; and urn 3 has 1 red and 6 blue chips. An urn is selected at random by tossing two fair coins and selecting the urn corresponding to the number of heads plus one. A chip is selected and it is red. What is the probability that the chip came from urn 2?

If we let A be the event of a red chip being selected and b_i, $i = 1, 2, 3$, the event of urn i being selected, then B_1, B_2, and B_3 form a partition of the sample space and we know the following probabilities:

$$P(B_1) = 1/4,\ P(B_2) = 1/2,\ P(B_3) = 1/4,\ P(A/B_1) = 3/7$$

$$P(A/B_2) = 4/7, \text{ and } P(A/B_3) = 1/7$$

Therefore,

$$P(B_1/A) = \frac{(3/7)(1/4)}{(3/7)(1/4) + (4/7)(1/2) + (1/7)(1/4)} = 1/4$$

$$P(B_2/A) = 2/3, \text{ and } P(B_3/A) = 1/12.$$

Often, the set B_i in the partition of the sample space are referred to as hypotheses. The assertion that $P(B_2/A)$ of the theorem is a statement about the probability of a hypothesis, given the occurrence of an event A. The probability $P(B_2)$ is called a priori (prior) probability about the hypothesis B_2 and the probability $P(B_2/A)$ is called posteriori (posterior) probability about the same hypothesis.

CHAPTER 5

RANDOM VARIABLES

5.1 DEFINITION OF A RANDOM VARIABLE

Many random experiments have a natural numerical description of the outcomes, such as the number of letters in a word or the number of dots on the uppermost face of a die. If a random experiment does not have a numeral description, we can give an assignment of a set of real numbers to represent the outcomes. Thus, to each elementary event in a sample space, we somehow assign a real number, thereby defining a function.

> Random Variable
>
> A function whose domain is a finite or countably infinite sample space S and whose codomain is the set of real numbers is called a random variable.

The usual notation is to let a capital letter, such as X or Y, represent the functional correspondence, $X(e_i)$ denote the numerical value of the random variable for the elementary event e_i, and let S_x denote the actual range set of the function X in the codomain of real numbers. For example, flip a coin 3 times in succession. This experiment is a sequence of three Bernoulli trials and the basic sample space S is the 8 3-tuples of H and T. That is,

$$S = \{(H,H,H), (H,H,T), (H,T,H), (T,H,H), (H,T,T), (T,H,T), (T,T,H), (T,T,T)\}$$

$$e_1 = (H,H,H), e_2 = (H,H,T), e_3 = (H,T,H), e_4\ (T,H,H), e_5 = (H,T,T), e_6 = (T,H,T), e_7\ (T,T,H), \text{ and } e_8 = (T,T,T)$$

a. Define the random variable X on S to be the number of heads in any outcome. Then

$$S_X = \{0, 1, 2, 3\}, \text{ and}$$

the correspondence for the sample space S is given by

$X(e_1) = 3, X(e_2) = 2, X(e_3) = 2, X(e_4) = 2, X(e_5) = 1, X(e_6) = 1$
$X(e_7) = 1$, and $X(e_8) = 0$.

b. Define the random variable Y on S to be +l if there are more H's than T's, and −1 if there are fewer H's than T's. Then

$$S_y = \{-1, +1\}$$

and $Y(e_1) = +1, Y(e_2) = +1, Y(e_3) = +1, Y(e_4) = +1, Y(e_5) = -1, Y(e_6) = -1, Y(e_7) = -1$, and $Y(e_8) = -1$

In general,

> If X is a random variable whose domain is the sample space S with probability function P, and if S_x is the range set of real numbers for X with
>
> $$S_X = \{x_1, x_2, x_3, \ldots, x_n\}$$
>
> then the function P_x defined for any elementary event $\{x_i\}$ by
>
> $$P_X(x_i) = P(\{e_j \mid e_j \varepsilon S, X(e_j) = x_i\})$$
>
> is a probability function for S_X
>
> If A is an arbitrary event of S_X then
>
> $$A = \{x_i \mid x_i \varepsilon S_X, a \le x_i \le b\}, a, b, \text{ are real numbers}$$
>
> and
>
> $$P_X(A) = \sum_{x_i \varepsilon A} P_X(x_i)$$

Remark

> If the sample space S_X is a finite set, the random variable X is called a finite random variable. When S_X is a countably infinite set, the random variable X is called a discrete random variable.

5.2 POINT AND DISTRIBUTION FUNCTIONS

5.2.1 PROBABILITY POINT FUNCTION

Let X be a random variable on a sample space S with probability function P, and let $S_X = \{x_1, x_2, x_3, \ldots,\}$ be the range of X. For any event $A \subseteq S_X$, the probability function P_X was defined in section 5.1 as

$$P_X(A) = \sum_{x_i \varepsilon A} P_X(x_i)$$

Define a function Q on S_X that coincides with P_X only at the point x_i; that is,

$$Q(x_i) = P_X(x_i)$$

Since Q is defined only at the points x_i of S_X and not on the events A (A is a subset of S_X), Q is called a probability point function. This function can be defined formally as follows.

> Probability Point Function
>
> The function Q defined for all real number $S_X = \{x_1, x_2, x_3, \ldots\}$ by
>
> $$Q(x_i) = P_X(x_i)$$
>
> is called the probability point function for the random variable X

If S_X is a finite sample space with n elements and Q is a probability point function, then

1. $0 \leq Q(x_i) \leq 1$ for all $x_i \varepsilon S_X$
2. $Q(x_i) = 1$

The function Q is sometimes called the frequency distribution of X, or the probability function of the discrete random variable X.

5.2.2 PROBABILITY DISTRIBUTION FUNCTION

Sometimes it is convenient to extend the definition of the probability point function Q defined above to the set of all real numbers by defining $Q(x) = 0$ if $x \notin S_X$. Thus, if

$$S_X = \{x_1, x_2, x_3, \ldots x_n\}$$

then

$$Q(x) = \begin{cases} x_i \text{ if } x = x_i, \text{ i} = 1, 2, 3, \ldots, n \\ 0 \text{ if } x \notin S_X \end{cases}$$

If A is an event of S_X, then

$$P_X(A) \sum_{x_i \varepsilon A} Q(x_i)$$

where the sum is taken over all points in A such that $Q(x_i) > 0$. The probabilities of events such as $a \le X \le b$, where a and b are arbitrary real numbers, can be written in terms of $Q(x)$ as

$$P_X(a \le X \le b) = \sum_{a \le x \le b} Q(x)$$

where the sum is taken over all x in the interval such that $Q(x) \ge 0$.

For a sample space S_X generated by a random variable X, a new probability function can be defined as follows.

> Probability Distribution Function
>
> The function F, whose value for each real number x is given by
>
> $$F(x) = P_X(X \le x)$$
>
> is called the probability distribution function for the random variable X.

The probability distribution function can be specified by the probability point function as follows:

$$F(x) = \sum_{x_i \le x} Q(x_i)$$

where the sum is taken over all points $x_i \le x$ such that $P(x_i) > 0$.

5.2.3 PROPERTIES OF THE DISTRIBUTION FUNCTION

Distribution functions have three important properties.

> 1. $F(x)$ is a nondecreasing function of x. That is, if u and v are real numbers such that $u < v$, the $F(u) \le F(v)$. If the sample space is $S_X = \{x_1, x_2, \ldots x_n\}$, where $x_1 < x_2 < x_3 < \ldots < x_n$, then
> 2. $F(x) = 0$ for $x < x_1$, and
> 3. $F(x) = 1$ for $x \ge x_n$.

For example, if $S_X = \{0, 1, 2, 3\}$, and Q is defined as

$$Q(x) = \begin{cases} \binom{3}{x} \times (1/2)^x \times (1/2)^{3-x} & \text{for } x = 0, 1, 2, 3 \\ 0, \text{ Otherwise} \end{cases}$$

then, for $x = 0$, $Q(x) = 1/8$; for $x = 1$, $Q(x) = 3(1/2)(1/2)^2 = 3/8$; and so forth.

5.3 CONCEPTS OF AVERAGE

5.3.1 ARITHMETIC AVERAGE (MEAN)

> Arithmetic Average
>
> Let $x_1, x_2, \ldots, x_n$ be the values of n items or observations. Then the arithmetic average of these items, denoted by x, is defined as
>
> $$\bar{x} = \frac{x_1 + x_2 + \ldots + x_n}{n} = \frac{\sum_{i=1}^{n} x_i}{n}$$
>
> The arithmetic average is also called the arithmetic mean.

Note that the arithmetic mean is that number which identifies the center of the observed values.

5.3.2 WEIGHTED MEAN

In averaging a set of observations, it is often necessary to compute the so-called weighted average to arrive at the desired measure of central location.

> Weighted Mean
>
> Let $x_1, x_2, \ldots x_n$ be a set of values of n items or observations, and let $x_{\cdot 1}, x_{\cdot 2}, \ldots, x_{\cdot n}$ be the distinct numbers in the set with weights or frequencies $w_1, w_2, \ldots w_k$, respectively. The weighted mean, denoted by $\bar{x}_w$, is defined as
>
> $$\bar{x}_w = \frac{\sum_{i=1}^{k} x_i w_i}{\sum_{i=1}^{k} w_i}$$

For example, if a final examination in a course is weighted 3 times as much as a quiz and a student has a final examination grade of 85 and quiz grades of 70 and 90, then the mean grade is given by

$$\bar{x}_w = \frac{(1)(70) + (1)(90) + (3)(85)}{1 + 1 + 3} = \frac{415}{5} = 83$$

5.3.3 AVERAGE OF A FUNCTION WITH RESPECT TO A SET OF NUMBERS

The idea of the averages of a set of numbers can be extended to obtain the average of a function with respect to a set of numbers.

> Average Of A Function With Respect To A Set Of Numbers
>
> Let $x_1, x_2, \ldots, x_n$ be a set of real numbers. Let $g(x)$ be a function whose domain contains the numbers $x_1, x_2, \ldots x_n$. The average of $g(x)$ with respect to the set of these numbers is denoted by $\overline{g(x)}$ and is defined as
>
> $$\bar{g}(\bar{x}) = \frac{1}{n}\sum_{i=1}^{n} g(x_i) = \sum_{j=1}^{k} g\left(x_j'\right)f\left(x_j'\right)$$
>
> where the numbers $x_1', x_2', \ldots, x_k'$ are the distinct numbers that appear in the set with the relative frequencies $f(x_1'), f(x_2'), \ldots, f(x_k')$ respectively.

5.4 MATHEMATICAL EXPECTATION OF DISCRETE RANDOM VARIABLES

Let X be a discrete random variable defined on a sample space $S = \{x_1, x_2, \ldots, x_n\}$, having a finite range $S_X = \{x_1, x_2, \ldots, x_n\}$ with point probabilities $Q(x_i) = 1/n$. The possible outcomes of the random experiment in S_X have average

$$\mu = \sum_{i=1}^{n} \frac{x_i}{n} = \sum_{i=1}^{n} x_i \frac{1}{n} = \sum_{i=1}^{n} x_i Q(x_i)$$

If some discrete model other than the equally likelihood model applies, then some values of X are more likely to occur than others. Then the equation defining, μ may be viewed as defining a weighted average of the outcomes, where each x_i is weighted by its point probability. This weighted average is called the mean of X or the expected value of X

> Expected Value Of A Discrete Random Variable
>
> Let X be a discrete random variable with range $S_X = \{x_1, x_2, \ldots, x_n\}$ and a probability point function Q. The expected value (or mean) of X, denoted by $E(X)$, is defined as
>
> $$E(X) = \sum_{i=1}^{n} x_i Q(x_i)$$

For example, let X be the first digit that results from a call to the random number generator in a computer system. Then $S_X = \{0, 1, 2, \ldots, 9\}$, and $Q(x_i) = 0.1$. Thus,

$$E(X) = \sum_{i=1}^{10} x_i Q(x_i) = (0)(.1) + (1)(.1) + \ldots + (9)(.1)$$
$$= 0.1(0 + 1 + 2 + \ldots + 9)$$
$$= 0.1(45)$$
$$= 4.5$$

Note that the expected value of X is not an outcome to be expected since an outcome for this experiment can never be 4.5.

5.4.1 PROPERTIES OF EXPECTED VALUES

1. Let X be a random variable defined on a sample space S, and let c be any constant. Then
   i) $E(cX) = cE(X)$
   ii) $E(X + c) = E(X) + c$
2. Let $X_1, X_2, \ldots, X_n$ be random variables defined on the same sample space S. Then

   $E(X_1 + X_2 + \ldots + X_n) = E(X_1) + E(X_2) + \ldots + E(X_n)$

5.4.2 CALCULATION OF EXPECTED VALUE

The expected value can be calculated explicitly for random variables with standard probability functions such as the following.

1. **Random Variables With Uniform Probability Functions**

If X is a random variable with $S_x = \{x_1, x_2, \ldots x_n\}$ and a uniform probability point function $Q(x_i) = 1/n$, then

$$E(X) = \frac{\sum_{i=1}^{n} x_i}{n}$$

2. **Bernoulli Random Variables**

Let X be a Bernoulli random variable defined on a sample space S, $S_X = \{0, 1\}$, and $Q(1) = p$, where p is the probability of success in a trial. Then

$$E(X) = P$$

3. **Binomial Random Variables**

Let X be a Binomial random variable and $S_x = x_1, x_2, \ldots, x_n$ be the number of successes in n trials with probability p for success on each trial. Then

$$E(X) = np$$

4. **Independent Random Variables**

If X and Y are independent random variables , then

$$E(XY) = E(X) \times E(Y)$$

5.4.3 EXPECTATION OF A FUNCTION OF A RANDOM VARIABLE

If X is a random variable, then X defines a function from a sample space S to the set of real numbers. If g is a real-values function defined over the set of real numbers, then the composition of X and g, denoted by $g(X)$, is also a function from the sample space S to the set of real numbers. Thus, $g(X)$ is itself a random variable. Let $Y = g(X)$. Then Y is called a function of the random variable X.

If X is a random variable with range $S_X = \{x_1, x_2, \ldots, x_n\}$, $E(X)$ $= \sum_{i=1}^{n} x_i\, QX(x_i)$, and $Y = g(X)$ is a random variable of X, then

$$E(Y) = E(g(X)) = \sum_{i=1}^{n} g(x_i) Q_X(x_i)$$

A special case of the function g occurs when g is the sum of two other functions g_1 and g_2. For example $Y = X^2 + X$ can be written as

$Y = (X)^2 + X$; so if $g_1(X) = X^2$ and $g_2(X) = X$, then $g(X) = g_1(X) + g_2(X)$.

Let X be a random variable defined on a sample space S, let g_1 and g_2 be two functions whose domains include S_X. If $Y = g_1(X) + g_2(X)$, then

$$E(Y) = E(g_1(X)) + E(g_2(X))$$

Some other properties for the expected value of a function of a random variable are the following.

1. If X is a finite random variable and a and b are arbitrary real numbers, then

$$E(aX + b) = aE(X) + b$$

2. If X is a finite random variable and a and b are arbitrary real numbers, then

i) $E(aX) = aE(X)$

ii) $E(X + b) = E(X) + b$

iii) $E(b) = b$

iv) $E(X - E(X)) = 0$

5.5 VARIANCE OF DISCRETE RANDOM VARIABLES

The mean or expected value of a random variable is useful in describing the center or balancing point of its distribution. Suppose, however, that the random variables X, Y, and Z have the same expected value. That is, $E(X) = E(Y) = E(Z)$, then the probability functions of these variables are not necessarily identical. A quantity which measures the spread or variability is based on the expected value of the "square deviations from the mean." This measure is called the variance.

Variance

Let X be a random variable defined on a sample space S with range $S_X = \{x_1, x_2, \ldots x_n\}$ a probability point function Q, and an expected value $E(X)$. Then the variance of the random variable X, denoted by $V(X)$, is defined as

$$V(X) = E((X - E(X))^2)$$

$$= \sum_{i=1}^{n} (x_i - E(X)^2)\, Q(x_i)$$

Since $V(X)$ is a function of squared deviations from the mean, it measures the variability in squared units of X. Frequently, it is desirable and useful to have a measure in the original units of X. To accomplish this, we take the positive square root of $V(X)$. That is, X and σ_X have the same units. Because of its importance and practicality, we define the square root of $V(X)$ formally.

Standard Deviation

The square root of the variance of a random variable X is called the standard deviation of X and denoted by

$$\sigma_X = \sqrt{V(X)}$$

For example, if X denotes the uppermost face when a fair die is tossed, then $E(X)$, $V(X)$, and σ_X can be calculated as follows.

$$E(X)=\sum_{i=1}^{6} x_i Q(x_i)=1(1/6)+2(1/6)+3(1/6)+4(1/6)+5(1/6)+6(1/6)$$
$$=3.5$$
$$V(X)=\sum_{i=1}^{6} (x_i-E(X))^2 Q(x_1)=\sum_{i=1}^{6} (x_i-3.5)^2 Q(x_i)=\frac{35}{12}$$

and

$$\sigma_X=\sqrt{V(X)}=\sqrt{\frac{35}{12}}$$

5.5.1 COMPUTING FORMULA FOR V(X)

If the mean $E(X)$ of a random variable is not an integer, then using the formula

$$V(X)=\sum_{i=1}^{n} (x_i-E(X))^2 Q(x_i)$$

to compute the variance of the random variable X is cumbersome. The following formula gives us an easier way to compute it.

> If X is a random variable with a variance $V(X)$, then
> $$V(X) = E(X^2) - (E(X))^2$$

5.5.2 PROPERTIES OF VARIANCE FOR A SINGLE RANDOM VARIABLE

> 1. If X is a finite random variable, and a and b are arbitrary real numbers, then
> $$V(aX + b) = a^2 V(X)$$
> 2. If X is a finite random variable, and a and b are arbitrary real numbers, then
>   i) $V(X + b) = V(X)$
>   ii) $V(aX) = a^2 V(X)$

3. If X is a random variable with a uniform probability point function for n values, then

$$V(X) = \frac{1}{n}\sum_{i=1}^{n} (x_1 - E(X))^2$$

$$= \frac{1}{n}\sum_{i=1}^{n} x_i^2 - (E(X))^2$$

4. If X is a Bernoulli random variable with $S_X = \{0, 1\}$, then

$$V(X) = pq$$

where $p = Q(1)$ and $q = 1 - p$

5. If X is a binomial random variable for n trials, where p is the probability of a success on each trial, then

$$V(X) = np(1 - p) = npq$$

where $p + q = 1$

5.5.3 STANDARDIZED RANDOM VARIABLES

For every random variable X with expected value $E(X)$ and standard deviation σ_X, a random variable Z can be defined in such a way that Z has the same type of probability function as X; but $E(Z) = 0$, and $V(Z) = 1$.

Standardized Random Variables

If X is any random variable with expected value $E(X)$ and standard deviation $\sigma_X > 0$, then the random variable Z defined as

$$Z = \frac{X - E(X)}{\sigma_X}$$

is called the standardized random variable corresponding to X

For example, if a random variable X has a mean, $E(X) = 50$, a variance $V(X) = 75$, and Z is the standardized random variable corresponding to X, then the value of Z which corresponds to a value of $X = 110$ is given by

$$Z = \frac{X - E(X)}{\sigma_X} = \frac{110 - 50}{\sqrt{75}} = 6.93$$

CHAPTER 6

SUMS OF RANDOM VARIABLES

6.1 JOINT RANDOM VARIABLES

6.1.1 FUNCTIONS OF TWO VARIABLES

Let A and B be two sets of real numbers, and let

$$A \times B = \{(x,y) \mid x \varepsilon A, y \varepsilon B\}$$

be the cartesian product of A and B. The correspondence (x,y) to z, where z is a real number, is a mathematical function, f, of two variables x and y with domain $A \times B$ and codomain the set of real numbers. The value of the function at the point (x, y) is written as $z = f(x,y)$.

6.1.2 JOINT PROBABILITY POINT FUNCTIONS

Let X and Y be two random variables defined on the same sample space S, where the probability function is P and where

$$S_X = x_1, x_2, \ldots, x_m, S_Y = \{y_1, y_2, \ldots, y_n\}$$

The function Q defined on the set $S_X \, x S_Y = \{(x_i, y_j) \mid x_i, \varepsilon S_X, y_j \varepsilon S_Y\}$ by

$$Q(x_i, y_j) = P(X = x_i, Y = y_j)$$

is called the joint probability point function for the variables X and Y, or the joint distribution of X and Y.

A double entry table as the following table demonstrates a convenient way to represent a joint probability point function.

	y_1	y_2	...	y_n	Sum
x_1	$Q(x_1, y_1)$	$Q(x_1, y_2)$	...	$Q(x_1, y_n)$	$Q_X(x_1)$
.	.	.	.	.	.
.	.	.	.	.	.
.	.	.	.	.	.
x_m	$Q(x_m, y_1)$	$Q(x, y_2)$	...	$Q(x_m, y_n)$	$Q_X(x_m)$
Sum	$Q_Y(y_1)$	$Q_Y(y_2)$	...	$Q_Y(y_n)$	

where Q_X and Q_Y are the probability point functions for the sets S_X and S_Y.

The notion of joint probability point function (joint distribution) can be extended to any finite number of random variables $X, Y, \ldots, Z$ as follows,

> Let $X, Y, \ldots, Z$ be random variables on the same sample space S where the probability function is P and where
>
> $$S_X = \{x_1, x_2, \ldots, x_m\}$$
> $$S_Y = \{y_1, y_2, \ldots, y_n\}$$
> $$\vdots \qquad \vdots$$
> $$S_Z = \{z_1, z_2, \ldots, z_n\}$$
>
> The function Q defined on the product set
>
> $$S_X x\, S_Y x \ldots x S_Z = (x, y, \ldots, z)\; x\, S_X,\; y, S_Y, \ldots, z\, S_Z$$
>
> as
>
> $$Q(x_i, x_j, \ldots, z_k) = P(X = x_i, Y = y_j, \ldots, Z = z_k)$$
>
> is called the joint probability point function for the random variables $X, Y, \ldots, Z$.

6.1.3 MARGINAL PROBABILITY POINT FUNCTIONS

> Marginal Probability Point Functions
>
> Let X and Y be two random variables with joint probability function given by $Q(x_i, y_j)$. Then the marginal probability point functions of X and Y are given by
>
> $$Q_X(x_i) = \sum_{i=1}^{n} Q(x_i y_j)$$
>
> and
>
> $$Q_Y(y_j) = \sum_{i=1}^{m} Q(x_i y_j)$$

6.1.4 EXAMPLE OF JOINT AND MARGINAL PROBABILITY POINT FUNCTIONS

Flip a coin 4 times in succession. The sample space S is given by

$S =$ $\{(H, H, H, H), (H, H, H, T), (H, H, T, H), (H, T, H, H), (T, H, H, H),$
$(H, H, T, T), (H, T, H, T), (H, T, T, H), (T, H, T, H), (T, T, H, H),$
$(T, H, H, T), (H, T, T, T), (T, H, T, T), (T, T, H, T), (T, T, T, H),$
$(T, T, T, T)\}$

Define the random variable X on S to be the number of heads in any outcome. Then $S_X = \{0, 1, 2, 3, 4\}$. Also define the random variable Y on S to be +1, if there are more H's than T's , −1 if there are fewer H's than T's. In this case, $S_Y = \{-1, 0, +1\}$.

The joint probability function for X and Y can be constructed by examining the events $X = i, Y = j$ of S for $i = 0, 1, 2, 3, 4$ and $j = -1, 0, +1$. The results are given in the following table.

	y_j			
x_i	−1	0	+1	$Q_X(x_i)$
0	1/16	0	0	1/16
1	4/16	0	0	4/16
2	0	6/16	0	6/16
3	0	0	4/16	4/16
4	0	0	1/16	1/16
$Q_Y(y_j)$	5/16	6/16	5/16	

6.2 INDEPENDENT RANDOM VARIABLES

Recall that two events A and B in a sample space S with probability P are defined as independent provided that

$$P(A \cap B) = P(A) \times P(B)$$

This idea of independence can be extended to random variables so the joint probability point function (or the joint distribution) of independent random variables can be formed simply by the product of the marginal probability point functions.

Independent Random Variables

Let X and Y be finite random variables on the same sample space S where the probability function is P, and with $S_X = \{x_1, x_2, \ldots x_m\}$

and $S_Y = \{y_1, y_2, \ldots, y_n\}$.
Then, the random variables X and Y are said to be independent if

$$P(X = x_i\ \ Y = y_j) = P(X = x_i) \times P(Y = y_j)$$

for all possible pairs of values of x_i and y_j. That is,

$$Q(x_i, y_j) = Q_X(x_i)\ Q_Y(y_j)$$

for the joint probability point functions.

For example, let X and Y be independent random variables with the following distributions:

x_i	0	1	2	3
$Q_X(x_i)$	.4	.3	.2	.1

y_j	0	1	2
$Q_Y(y_j)$	.5	.3	.2

Since X and Y are independent random variables, the joint probability point function can be computed using the expression

$$Q(x_i, y_j) = Q_X(x_i)\ Q_Y(y_j)$$

with results given in the following table.

	Y			
X	0	1	2	$Q_X(x_i)$
0	.20	.12	.08	.40
1	.15	.09	.06	.30
2	.10	.06	.04	.20
3	.05	.03	.02	.10
$Q_Y(y_j)$	.50	.30	.20	

The notion of independence of random variables can be extended to any finite number of random variables $X, Y, \ldots, Z$ as follows.

Mutually Independent Random Variables

Let $X, Y, \ldots, Z$ be finite random variables defined on the same sample space S where the probability function is P and with

$$S_X = \{x_1, x_2, \ldots x_m\}$$
$$S_Y = \{y_1, y_2, \ldots y_n\}$$
$$S_Z = \{z_1, z_2, \ldots z_n\}$$

Then, the random variables $X, Y, \ldots, Z$ are mutually independent if

$$P(X = x_i, Y = y_j, Z = Z_k) = P(X = x_i) \times P(Y = y_j) \ldots P(Z = z_k)$$

for any values x_i, y_j, z_k and for all possible subsets of the set of *n* random variables.

6.3 SUMS AND PRODUCTS OF RANDOM VARIABLES

6.3.1 PROBABILITY POINT FUNCTIONS

If X and Y are random variables defined on the same sample space, then $X + Y$ and XY are also random variables defined on S.

Probability Point Functions of $X + Y$ and XY

Let X and Y be random variables defined on the same sample space S, and let $Z = X + Y$ and $W = XY$, with

$$S_X = \{x_1, x_2, \ldots, x_m\}$$

$$S_Y = \{y_1, y_2, \ldots, y_n\}$$

$$S_Z = \{z_1, z_2, \ldots, z_t\}$$

$$S_W = \{w_1, w_2, \ldots, w_h\}$$

Then

a. the probability point function for $Z = X + Y$ is given by

$$Q_Z(z_h) = \sum_{x_i + y_j = Z_k} Q(x_i, y_j)$$

for $k = 1, 2, \ldots t$, where the sum is taken over all pairs (x_i, y_j) such that $x_i + y_j = z_k$. The probability point function for $W = XY$ is given by

$$Q_w(w_r) = \sum_{x_i + y_j = w_r} Q(x_i, y_j)$$

for $r = 1, 2, 3, \ldots, h$, where the sum is taken over all pairs (x_i, y_j) such that $x_i y_j = w_r$

For example, if the joint probability for the random variables X and Y is given as shown in the following table.

	Y		
X	0	1	2
1	.15	.05	.05
2	.10	.05	.10

3	.10	.05	.10
4	.10	.05	.10

and $Z = X + Y$, then $S_Z = \{1, 2, 3, 4, 5, 6,\}$

$$P(Z = 3) = P(X = 3, Y = 0) + P(X = 2, Y = 1) + P(X = 1, Y = 2)$$
$$= 0.10 + 0.05 + 0.05$$
$$= 0.20$$

The complete probability point function Q_Z is given as

Z_k	1	2	3	4	5	6
$Q_z(z_k)$	.15	.15	.20	.25	.15	.10

If $W = XY$, then $S_W = \{0, 1, 2, 3, 4, 6, 8\}$, and the values for Q_W are calculated using the equation

$$Q_W(w_r) = \sum_{x_i y_j = w_r} Q(x_i, y_j)$$

The complete probability point function is

w_r	0	1	2	3	4	6	8
$Q_W(w_r)$	.45	.05	.10	.05	.15	.10	.10

where for example

$$P(W = 2) = P(X = 2, y = 1) + P(X = 1, y = 2) = 0.05 + 0.05 = 0.10$$

6.3.2 EXPECTATION OF SUMS AND PRODUCTS OF RANDOM VARIABLES

Let X and Y be random variables defined on the same sample space S with expected values $E(X)$ and $E(Y)$. Then

a. The expected value of $X + Y$ is given by

$$E(X + Y) = E(X) + E(Y)$$

b. If in addition the random variables X and Y are independent, then

$$E(XY) = E(X) \times E(Y)$$

The notion of expectation of the sum of two random variables can be extended to the sum of an arbitrary number of random variables as follows.

Let $x_1, x_2, \ldots, x_n$ be n random variables defined on the same sample space S with $E(X_j)$, $j = 1, 2, \ldots, n$, then

a. $$E\left(\sum_{j=1}^{n} X_j\right) = \sum_{j=1}^{n} E(X_j).$$

That is, the expected value of the sum is equal to the sum of the expected values.

b. $$E\left(\sum_{j=1}^{n} a_j X_j\right) = \sum_{j=1}^{n} a_j E(X_j)$$ where the a_j's are constants.

6.3.3 COVARIANCE OF TWO RANDOM VARIABLES

In studying the behavior of two random variables simultaneously, we are faced with trying to find a measure of how the random variables vary jointly. That is, we are interested in the study of the concept of covariance.

Covariance Of Two Random Variables

Let X and Y be random variables defined on the same sample space S. Then the covariance of X and Y is the expected value of the product of the deviations $X - E(X)$ and $Y - E(Y)$; that is,

$$\text{Cov}(X, Y) = \sqrt{xy} = E\{[X - E(X)][Y - E(Y)]\}$$

It is clear that $\text{Cov}(X, Y) = \text{Cov}(Y, X)$, and the covariance of a random variable with itself is its variance because

$$\begin{aligned}\text{Cov}(X, X) &= E\{[X - E(X)][X - E(X)]\}\\ &= E[(X - E(X))^2]\\ &= V(X)\end{aligned}$$

The following computational formula for the covariance is useful

Computational Formula For Covariance

$$\text{Cov}(X, Y) = E(XY) - E(X)E(Y)$$

Thus, if X and Y are independent random variables, then

$$\begin{aligned}\text{Cov}(X, Y) &= E(XY) - E(X)E(Y)\\ &= 0\end{aligned}$$

For example, given the joint probability point functions of two random variables X and Y and the respective marginal probability point functions Q_X and Q_Y,

		X				
Y	1	2	3	4	5	$Q_Y(y_j)$
1	.10	0	0	0	0	.10
2	0	.20	0	.10	0	.30
3	0	0	.20	0	0	.20
4	0	.10	0	.20	0	.30
5	0	0	0	0	.10	.10
$Q_X(x_i)$	.10	.30	.20	.30	.10	

then

$$E(X) = \sum_{i=1}^{5} x_i Q_X(x_i) = 1(.10) + 2(.30) + 3(.20) + 4(.30) + 5(.10)$$

$$= 3$$

$$E(Y) = \sum_{i=1}^{5} y_j Q_Y(y_j) = 3$$

$$E(XY) = 1(.10) + 4(.20) + 9(.20) + 16(.20) + 25(.10) + 8(.10) + 8(.10) = 10$$

Thus,

$$\text{Cov}(X, Y) = E(XY) - E(X)E(Y)$$

$$= 10 - 3(3)$$

$$= 1$$

6.4 VARIANCE OF SUMS OF RANDOM VARIABLES

Variance is one of the most important and useful properties of independent random variables.

> Let X and Y be random variables defined on the same sample space S. If X and Y are independent, then
>
> $$V(X + Y) = V(X) + V(Y)$$

For example, toss a fair die twice. Let X be the random variable whose values are the number of spots on the first toss, and Y be the random variable whose values are the number of spots on the second toss. Then,

$$E(X) = \sum_{i=1}^{6} x_i Q_X(x_i) = 3.5$$

$$V(X) = \sum_{i=1}^{6} (x_i - E(X))^2 Q_X(x_i) = 12/35$$

$$E(Y) = E(X) = 3.5, \text{ and } V(Y) = 35/12$$

Hence, $V(X + Y) = V(X) + V(Y) = 35/12 + 35/12 = 35/6$.

The additive property of the variance can be generalized as follows.

> Let $x_1, x_2, \ldots, x_n$ be n mutually independent random variables defined on the same sample space S, with variances $V(X_i)$, $i = 1, 2, \ldots, n$.
>
> Then
>
> $$V\left(\sum_{i=1}^{n} X_i\right) = \sum_{i=1}^{n} V(x_i)$$

6.5 SAMPLE RANDOM VARIABLES–SAMPLE MEAN

6.5.1 SAMPLE RANDOM VARIABLES

Suppose a box contains N balls each having a number on it. The same number may appear on several balls. Let $Y_1, Y_2, \ldots, Y_k = N$ be the distinct numbers on the balls. If n_j is the number of balls with Y_j, then the relative frequency of proportion of balls having the number Y_j is n_j/N. Note that

$$n_j = N, \text{ and } \frac{n_j}{N} = 1$$

If one ball is selected at random and we let Y be the number on the ball, then the probability function P of Y is,

$$P_Y(y_j) = \frac{n_j}{N}$$

Suppose we select n balls with replacement, then we can visualize this as a n-stage experiment and the assignment of probabilities is made using the product rule for independent events.

Let Y_i be a random variable whose value is the number on the ith ball drawn for $i = 1, 2, \ldots, n$. Then, for the n-stage experiment, each Y_i has the same probability point function. This set of random variables is called a random sample from the population Y.

> Let Y be a random variable and let $Y_1, Y_2, \ldots, Y_n$ denote repeated observations of Y. Each of these Y_i is a random variable whose probability point function is the same as Y, and the set $Y_1, Y_2, \ldots Y_n$ is a set of mutually independent and identically distributed sample variables of size n of the random variable Y.

6.5.2 SAMPLE MEAN AND VARIANCE

> Let $X_1, X_2, \ldots, X_n$ be n mutually independent, identically distributed random variables of the random variable X whose mean, μX and variance σ^2_x. Then the sample mean is the random variable
>
> $$\frac{\sigma_X^2}{X} = \frac{1}{n}\sum_{i=1}^{n} x_i$$
>
> and the sample variance is the random variable
>
> $$\sigma_X^2 = \sum_{i=1}^{n} \frac{(x_i - \bar{x})^2}{n-1}$$

Since X is a random variable defined on a sample space and $x_1, x_2, \ldots, x_n$ are defined on the same sample space, we can use the concepts of sums and products of random variables discussed in the previous two sections to define the expectation and variance of the random variable.

> Let n be a positive integer, X a random variable for which expectation and variance exist, and let, $\overline{X}$ be the sample mean for X for a sample of size n. Then
>
> $$V(\overline{X}) = \frac{1}{n}V(X)$$
>
> and
>
> $$V(\overline{X}) = \frac{1}{n}V(X)$$

6.6 CHEBYSHEV'S INEQUALITY AND THE LAW OF LARGE NUMBERS

6.6.1 CHEBYSHEV'S INEQUALITY

To discuss the law of large numbers, we first need an important inequality called the Chebyshev's inequality.

> Chebyshev's Inequality
> Let X be a random variable with expected value $\mu_X = E(X)$, and let $\varepsilon > 0$ be any positive real number, then
>
> $$P\left(|X - \mu_X| \geq \varepsilon\right) \leq \frac{1}{\varepsilon^2} V(X)$$
>
> or
>
> $$P\left(|X - E(X)| \geq \varepsilon\right) \leq \frac{V(X)}{\varepsilon^2}$$

In words, this says that X differs from $\mu_X = E(X)$ by more than ε with probability at most $V(X)/\varepsilon^2$. The larger ε is, the less likely that $|X - \mu_X| > \varepsilon$. For example, let X be a random variable with probability point function as follows.

X	0	1	2
$Q_X(x)$	5/16	8/16	3/16

Then $E(X) = 7/8$, and $V(X) = 0.4844$ and the probability that X is at or within

a. One standard deviation of the mean is

$$P\left\{|X - 7/8| \leq \sqrt{0.4844}\right\} = P\{X - 7/8| \leq 0.696\} = \frac{8}{16} =.05$$

b. Two standard deviations of the mean is

$$P\left\{|X - 7/8| \leq 2\sqrt{0.4844}\right\} = P\left\{|X - 7/8| \leq 2(0.696)\right\} = 1$$

6.6.2 LAW OF LARGE NUMBERS FOR DISCRETE RANDOM VARIABLES

Let X be a random variable with mean, $\mu_X = E(X)$ and variance $V(X)$. If $x_1, x_2, x_3, \ldots, x_n$ is a random sample of size n of random variables of X with sample mean $\mu_{\bar{X}} = E(X)$ and $\varepsilon > 0$ is a real number, then

$$P\left\{\left|\mu_X - \mu_{\overline{X}}\right| \geq \varepsilon\right\} \to 0 \text{ as } n \to \infty$$

Equivalently,

$$P\left\{\left|\mu_X - \mu_{\overline{X}}\right| < \varepsilon\right\} \to 1 \text{ as } n \to \infty$$

That is,

$$P(\mu_X - \varepsilon) \leq \mu_{\overline{X}} \leq u_x + \varepsilon$$

approaches 1 as *n* increases without bound.

CHAPTER 7

COUNTABLY INFINITE RANDOM VARIABLES

7.1 COUNTABLY INFINITE PROBABILITY SPACES

If a sample space has an infinite number of elements (or points), then the way that a probability function is defined depends upon whether or not the sample space is countable. A sample space is countably infinite if the elements can be counted; that is, the elements can be put in one-to-one correspondence with the set of natural numbers, and noncountably infinite otherwise.

> If S is a countably infinite sample space, then we write
>
> $$S = \{e_1, e_2, \ldots, e_n, \ldots\}$$
>
> to indicate all the elements of S, and the collection $E_i = \{e_i\}$, $i = 1, 2, 3, \ldots$ to represent the collection of all elementary events of S. As in the finite sample space case, the set of all subsets of S comprises the set E of all events.
>
> The sample space S and this set E together with a set function P whose domain is E and whose range is the set of real numbers satisfy both sets of axioms of a probability space stated in section 2.1.

7.2 COUNTABLY INFINITE PROBABILITY FUNCTIONS

If $\{e_1, e_2, \ldots, e_n, \ldots\}$ is a countably infinite sample space, then the probability function P is defined exactly in the same way as in the finite case, except that the sum must now be a convergent infinite sum; that is, P is defined for elementary events $E_i = \{e_i\}$, $i = 1, 2, 3, \ldots$, and is then extended to an arbitrary event $A \subset E$ by adding the probabilities for the elementary events in A.

If $P(E_i) = p_i$, i = 1, 2, 3,..., where $0 \le p_i \le 1$ for all i, and

$$\sum_{i=1}^{\infty} p_i = 1$$

then P defines the probability function for the countably infinite sample space S. If A is any event, then

$$P(A) = \sum_{e_{ij} \varepsilon A} P(E_{ij})$$

where the sum is taken over all e_{ij} εA

Note that the expression $\sum_{i=1}^{\infty}$ is an infinite series. While the expression

$\sum_{e_{ij} \varepsilon A}$ may or may not be an infinite series.

For example, a coin is bent so that the probability of a head coming up is 3/4 . The coin is tossed until a head occurs. Find the probability that the coin was tossed less than 10 times.

Let S = $\{e_1, e_2, \ldots, e_n, \ldots\}$ where $e_1 = H$, $e_2 = TH$, $e_3 = TTH$, ... , Then

$$P(E_1) = 3/4 \text{ , } P(E_2) = (1/4)(3/4), P(E_3) = (1/4)(1/4)(3/4)$$

In general,

$$P(E_k) = (1/4)^{k-1} (3/4)$$

The event A "the coin is tossed less than 10 times" is

$$A = \{e_1, e_2, \ldots e_9,\}$$

$$P(A) = \sum_{k=1}^{9} P(E_k) = \sum_{k=1}^{9} (1/4)^{9-1}(3/4) = \frac{3}{4} \times \frac{1-\left(\frac{1}{4}\right)^9}{1-\frac{1}{4}}$$

$$= 1 - \left(\frac{1}{4}\right)^9$$

Note that the infinite series, $\sum_{i=1}^{\infty} ar^{k-1}$ is an infinite geometric series. In our example $a = 3/4$, $r = 1/4$. This implies that

$$\sum_{i=1}^{\infty} (1/4)^{k-1}(3/4) = \frac{3/4}{1-1/4} = 1$$

7.3 COUNTABLY INFINITE RANDOM VARIABLES

7.3.1 PROBABILITY POINT AND DISTRIBUTION FUNCTIONS

Random variables, as defined in chapter 5, apply with only a slight change to the countably infinite sample space. The domain of a random variable is now an infinite set instead of a finite set of outcomes.

The two functions closely associated with a random variable are the probability point function and the probability distribution function.

> Probability Point Function and Probability Distribution Function
>
> A probability point function Q defined on a countably infinite sample space
>
> $$S = \{x_1, x_2, \ldots, s_n, \ldots\}$$
>
> is a function with the properties:
>
> a. $0 \leq Q(x_i) \leq 1$ for all i
>
> b. $\sum_{i=1}^{\infty} Q(x_i) = 1$
>
> The probability distribution function is
>
> $$F(x) = \sum_{X_n < X} Q(x_n)$$

For example, a coin is tossed until the first time that a head turns up. Let the outcome of the experiment , x, be the first time that a head turns up. Then the possible outcomes of our experiment are

$$S = \{1, 2, 3, \ldots\}$$

The probability that a head turns up on the first toss is 1/2, and the probability that a head turns up on the second toss is the probability that a tail turns up on the first toss and followed by a head on the second toss . This probability is equal to 1/4 . The probability that we have tails followed by a head is 1/8 and so forth.

Thus, $Q(n) = \frac{1}{2^n}$, for $n = 1, 2, 3, \ldots$, and

$$\sum_{i=1}^{\infty} Q(x_i) = 1/2 + 1/4 + 1/8 + \ldots = 1,$$

Since 1/2 + 1/4 + 1/8 + ... is the sum of a geometric series.

Now, if we let A be the event that the first time a head turns up is after an even number of tosses, then $A = \{2, 4, 6, 8, \ldots,\}$ and

$$P(A) = 1/4 + 1/16 + 1/64 + \ldots \frac{1/4}{1-1/4} = 1/3$$

If B is the event that a head turns up after an odd number of tosses, then

$$F(B) = 2/3$$

Thus, the probability distribution function F is given by

$$F(x) = \begin{cases} 1/3, & \text{if } x \text{ is even} \\ 2/3, & \text{if } x \text{ is odd} \end{cases}$$

where x represents the number of tosses that the first time a head turns up.

7.3.2 EXPECTATION AND VARIANCE OF COUNTABLY INFINITE RANDOM VARIABLES

The expected value and variance of a finite random variable are finite sums. For a countably infinite random variable, we obtain an infinite series.

> Expected Value, $E(X)$, and Variance, $V(X)$
>
> Let X be a random variable defined on a countably infinite sample S with probability point function Q. Then
>
> $$E(X) = \sum_{i=1}^{\infty} x_i \times Q(x_i)$$
>
> $$V(X) = \sum_{i=1}^{\infty} (x_i - E(X))^2 \times Q(x_i)$$
>
> provided that these infinite series converge absolutely.

7.4 GEOMETRIC RANDOM VARIABLES

Consider a Bernoulli trials experiment continued for an infinite number of trials. For example, a coin is tossed an infinite number of times. Let X be the random variable whose values are the number of repetitions up to and including the first success. Let p be the probability for success on each trial and $q = 1-p$ be the probability of failure. Then,

$S = \{1, 2, 3, \ldots\}$ with $x_n = n$ and $Q(x_n) = (1-p)^{n-1}p$, for $n = 1, 2, 3, \ldots$ Note that $0 = Q(x_n) = 1$, and

$$\sum_{n=1}^{\infty} Q(x_n) = 1$$

Hence, the function $Q(n) = (1-p)^{(n-1)} p$ is a probability point function. But

$$\sum_{n=1}^{\infty} (1-p)^{(n-1)} p = \sum_{n=1}^{\infty} q^{(n-1)} \times p$$ is a geometric series with $a = p$, and $r = q$.

> Geometric Probability Function.
>
> If a random experiment consists of Bernoulli trials with p the probability of success on each trial, then the probability point function for the sample space S of the number, n, of repetitions up to and including the first success is given by
>
> $$Q(n) = (1-p)^{(n-1)} p$$
>
> The function Q is called the geometric probability point function, and a random variable X with a geometric probability point function, with p as the probability of success on each trial is called a geometric random variable with parameter p.

7.5 EXPECTATION, *E(X)*, AND VARIANCE *V(X)* OF GEOMETRIC RANDOM VARIABLES

If X is a geometric random variable with parameter p, then

$$E(X) = \frac{1}{p}$$

and

$$V(X) = \frac{1-p}{p^2}$$

For example, suppose that the relative frequency of a particular blood type is 0.01. A search is made for a person with this particular blood type. If we let X be the trial number on which the correct blood type appears, then X is a geometric random variable with parameter $p = 0.01$. Thus, the expected value of X is

$$E(X) = 1/p = 1/(.01) = 100$$

and

$$V(X) = (1-p)/(p^2) = (1-.01)/(.01)^2 = 9900$$

7.6 THE POISSON RANDOM PROCESS

In applications, it is frequently desired to devise probability models for phenomena in which occurrences of events are observed over time. For example, it may be desired to model traffic density at a certain intersection in terms of the occurrences of automobile arrivals at the intersection over time or to model the arrival, at a Geiger counter, of subatomic particles that result from the radioactive decay of some material. One particular type of model is known as the Poisson Process.

> The Poisson Process
>
> A process with the following properties is called a Poisson process.
>
> 1. There are no arrivals at time zero.
> 2. The number of arrivals in nonoverlapping time intervals are independent.
> 3. The distribution of the number of arrivals depends only on the duration of the time interval.
> 4. For small time intervals, the probability of an arrival is proportional to the duration of the interval. The constant of proportionality is denoted by λ.
> 5. There are no simultaneous arrivals.

7.6.1 THE POISSON PROBABILITY POINT FUNCTION

Let X be a binomial random variable with parameters n and p, where n is a fixed number of Bernoulli trials and p is the probability of success in each trial. The sample space for such random phenomena is the countably infinite space $S = \{0, 1, 2, \dots\}$. As $n \to \infty$, and $p \to 0$, such that p is proportional to $1/n$; that is, for each n, the probability p depends on n and in such a way $np = \lambda$. With these results in mind,

$$P(X = x) = \binom{n}{x} p^x (1-p)^{(n-x)}$$

where x is the number of successes in n trials. For a fixed x, as $n \to \infty$, we obtain,

$$\lim_{n \to \infty} P(X = x) = \frac{\lambda^x}{x!} e^{-\lambda}$$

Thus, the probability point function of the random variable X is the function

$$Q(x) = P(X = x) = \frac{\lambda^x}{x!} e^{-\lambda}$$

Poisson Random Variables

If X is a random variable with S_x = {0, 1, 2, 3, ...} , and probability point function

$$Q(x) = \frac{\lambda^x}{x!} e^{-\lambda}$$

where λ is a positive constant, then X is called a Poisson random variable.

7.6.2 PROPERTIES OF THE POISSON RANDOM VARIABLE

If X is a Poisson random variable with parameter λ, then

1. The expected value of X is given by $E(X) = \lambda$.
2. The variance of X is given by $V(X) = \lambda$.

For example, at a supermarket, customers arrive at a checkout counter at the rate of 30 per hour. To find the probability that 5 or fewer customers will arrive in any 20-minute period, we note that the arrival of a customer at a checkout counter is the occurrence of random event S in the continuum of time so that a Poisson model is appropriate.

From the data, the mean rate at which the arrivals occur is 30 per hour, which is equivalent to 10 per each unit of 20-minutes duration. Thus, λ = 10, and $P(X = 5)$, where X is a Poisson random variable is

$$P(X = 5) = \sum_{x=0}^{5} \frac{(10)^x}{x!} e^{-10}$$
$$= 0.067$$

CHAPTER 8

CONTINUOUS RANDOM VARIABLES

8.1 SAMPLE SPACES OF REAL NUMBERS

If a piece of electronic equipment, such as a transistor, were put on test by operating until it failed, then the operating time to failure would be a random variable X. Even under the most controlled manufacturing conditions, different transistors would fail at different times. Usually, it would be preferred to allow for measurements of time in tenths of an hour, hundredths of an hour, and possibly even finer time divisions. Although there may not be a measuring device capable of reading it, conceptually, the true time to failure could be any positive real number, rational or irrational. Thus, the possible values of X range from 0 to ∞. That is, the sample space of X is $S = [0, \infty]$. The set $a \leq X \leq b$ is an event in S.

> The collection of all possible outcomes of a random experiment is called the sample space. For random experiments where the collection of outcomes is noncountably infinite, the sample space consists of intervals.

8.2 PROBABILITY DISTRIBUTION FUNCTIONS

> Let X be a random variable defined on a noncountably infinite sample space. The probability distribution function of X is the function F_X whose domain is the set of real numbers and whose codomain is also the set of real numbers defined by
>
> $$F_X(x) = P_X(X \leq x), \text{ where } X \text{ is a real number}$$
>
> In terms of F_X the probability $P_X(A)$ of an interval of real numbers $A = [a, b] \subset R$ (Real numbers) is given by
>
> $$P_X(a, b) = F_X(b) - F_X(a)$$

The real number x defines an event of S and the set $A = \{y \mid y \varepsilon S_x, y \leq x\}$. If S_x is the set of all real numbers, then the set A becomes the interval $(-\infty, x)$, and the event A has a probability number $F_X(x)$ attached to it. Thus, the domain of F_x is the interval [0, 1].

> Continuous Probability Distribution
> If $F_X(x)$ is a continuous function of x, then F_X is said to define a continuous probability distribution and X is called a continuous random variable.

8.3 PROBABILITY DENSITY FUNCTIONS

Let X be a random variable with probability distribution function F_X. A probability density function is a function f_X of X whose domain is the set R and whose range is also the set R; that is, f_X takes R into R such that

$$F_X(x) = \int_{-\infty}^{x} f_x(u)du$$

The probability density function f_X has the following properties:

1. $f_X(x) \geq 0$ for all x
2. f_X has at most a finite number of discontinuities in every finite interval of the real line.
3. $\int_{-\infty}^{\infty} f_X(u)du = 1$
4. For every interval [a, b],

$$P_X(a \leq X \leq b) = \int_{\alpha}^{b} f_X(x)\, dx$$

Thus, if X is a random variable defined on a noncountable infinite sample space S, then the range of X, together with the probability density function f_X form a continuous probability space. For example, a number is chosen at random from [0, 1] with uniform probability, and then this number is squared. To find the distribution of the square and its density, let $S =$ [0, 1] with uniform density function $Q(X) = 1$, let $x \varepsilon S$ be the number chosen, and let $X = x^2$ be its square. Then X is a random variable with domain $S =$ [0, 1] and range equals to [0, 1].

The probability distribution function F_X of X is given by

$$F_X(x) = P_X\{X \leq x\} = P_X\{x^2 \leq x\}$$
$$= P_X(X \leq \sqrt{x})$$

Thus,

$$F_X(x) = \begin{cases} 0 \text{ if } x \le 0 \\ x \text{ if } 0 < x < 1 \\ 1 \text{ if } 1 \le x \end{cases}$$

The corresponding density function f_X is given by

$$f_X(x) = \frac{d}{dx} F_X(x) = \left\{ \begin{matrix} 0 \\ \dfrac{1}{2\sqrt{\dfrac{x}{0}}} \end{matrix} \right\} \quad \begin{matrix} 0 \text{ if } x \le 0 \\ x \text{ if } 0 < x < 1 \\ 1 \text{ if } 1 \le x \end{matrix}$$

Note that F_X is continuous, but f_X is not.

8.4 FUNCTIONS OF A RANDOM VARIABLE

Let X and Y be random variables defined on the same sample space S, with probability distribution functions F_X and F_Y. Suppose that $Y = \phi(X)$ for some continuous function f that is strictly monotone on the range of X. Then F_Y and F_X are related by

$$F_Y(y) = F_X(\phi^{-1}(y))$$

Suppose, for example, that $Y = X^3$. Then $\phi(x) = x^3$, is strictly increasing for all x, and $\phi^{-1}(y) = y^{1/3}$. Then we have

$$F_Y(y)\ F_X(y^{1/3})$$

For example, assume that X has a uniform distribution on the interval [0, 1] and $Y = X^3$. Then since $F_X(x) = x$ if $0 \le x \le 1$, the distribution of Y is given by

$$F_Y(y) = \begin{cases} 0 & \text{if } y < 0 \\ y^{1/3} & \text{if } 0 \le y \le 1 \\ 1 & \text{if } y > 1 \end{cases}$$

The relation between the probability density functions f_X and f_Y follows by differentiation.

If ϕ is differentiable, and $y = \phi(x)$, then by the chain rule

$$F_Y(y) = \frac{d}{dy} F_Y(y) = \frac{d}{dy} F_X\left(\phi^{-1}(y)\right)$$

$$\frac{d}{d\left(\phi^{-1}(y)\right)} F_X\left(\phi^{-1}(y)\right) \times \frac{d}{dy}\left(\phi^{-1}(y)\right)$$

If ϕ is differentiable, it follows that $\phi^{-1}$is differentiable, and hence

$$f_Y(y)=\begin{cases} f_x\left(\phi^{-1}(y)\right)\times\frac{d}{dy}\left(\phi^{-1}(y)\right), \text{ if } y=\phi(x) \text{ for some } x \\ 0, \text{ if } y\neq\phi(x), \text{ for all } x \end{cases}$$

For example, if $Y = X^3$, then $\phi(x) = x^3$, $\phi^{-1}(x) = y^{1/3}$, and

$$\frac{d}{dy}\left(\phi^{-1}(y)\right)=\frac{1}{3}y^{-2/3}$$

Therefore,

$$f_y(y)=\begin{cases} f_x\left(y^{1/3}\right)\times\frac{1}{3}y^{-2/3}, \text{ if } 0\leq y\leq 1 \\ 0, \text{ otherwise} \end{cases}$$

A similar result holds if the function $\phi(x)$ is strictly decreasing on the range of X, so that $\phi(x_1) > (x_2)$ if $x_1 < x_2$ there. Then ϕ reverses inequalities and we have

$$\begin{aligned} P_Y(y) = P_Y(Y \leq y) &= P_Y(\phi(x) \leq y) \\ &= P_X(X \geq \phi^{-1}(y)) \\ &= 1 - P_X(X \leq \phi^{-1}(y)) \\ &= 1 - F_X(\phi^{-1}(y)) \end{aligned}$$

and it follows

$$f_Y(y) = -fx(\phi^{-1}(y)) \times \frac{d}{dy}(\phi^{-1}(y))$$

8.5 EXPECTATION, *E(X)*, AND VARIANCE, *V(X)*, OF A CONTINUOUS RANDOM VARIABLE *X*

8.5.1 EXPECTED VALUE, *E(X)*

Let X be a random variable defined on a continuous sample space S with probability density function f_x. Then the expected value or mean of X, written as μ_X or $E(X)$, is defined as follows

$$\mu_X = E(X) = \int_{-\infty}^{\infty} x f_X(x)dx$$

provided that $\int_{-\infty}^{\infty} x f_X(x)dx$ converges absolutely.

If $\int_{-\infty}^{\infty} x f_X(x)dx$ does not converge, then $E(X)$ is said not to exist.

$E(X)$ is interpreted as the value that we should expect to obtain if we perform a large number of independent experiments and average the resulting values of X. For example, suppose that X is a continuous random variable with probability density function f_X defined as follows:

$$f_X(x) = \begin{cases} 2x, & 0 < x < 1 \\ 0, & \text{otherwise} \end{cases}$$

then

$$E(X) = \int_{-\infty}^{\infty} x f_X(x)dx = \int_0^1 x f_X(x)dx = \int_0^1 x(2x)dx = \int_0^1 2x^2 dx = 2/3$$

8.5.2 PROPERTIES OF EXPECTATION

1. Let X be a continuous random variable with probability density function f_X and let $Y = \phi(x)$ where ϕ is a continuous real-values function. If $E(Y)$ exists, then

$$E(Y) = E[\phi(X)] = \int_{-\infty}^{\infty} \phi(x) f_x(x)dx$$

2. Let X and Y be continuous random variables defined on the same sample space S such that $E(X)$ and $E(Y)$ exist. If a, b, and c are arbitrary real numbers, then
   a. $E(X + Y) = E(X) + E(Y)$
   b. $E(cX) = cE(X)$
   c. $E(aX + b) = aE(X) + b$
3. If X and Y are independent continuous random variables defined on the same sample space S, then

$$E(XY) = E(X)E(Y)$$

More generally,

If $x_1, x_2, x_3, \ldots, x_n$ are n continuous random variables defined on the same sample space such that $E(X_i)$, $i = 1,2,3, \ldots, n$ exists, and $c_1, c_2, \ldots, c_n$ are arbitrary real numbers, then

$$E(c_1 x_1 + c_2 x_2 + \ldots + c_n X_n) = c_1 E(x_1) + c_2 E(x_2) + \ldots + c_n E(X_n)$$

8.5.3 VARIANCE, *V(X)*

Let X be a continuous random variable with probability density function f_X. The variance, σ^2_X or $V(X)$, is defined by

$$V(X) = E((X - E(X))^2)$$

$$= \int_{-\infty}^{\infty} (x - E(X)^2) \times f_X(x)dx$$

The positive square root of $V(X)$ is called the standard deviation σ_X; that is,

$$\sigma_X = \sqrt{V(X)}$$

We can interpret $V(X)$ as the value we should expect to obtain if we average the resulting values of $(X - \mu_X)^2$, the square of the deviations of X from μ_X. Note that $V(X)$ exists if and only if $E(X)$ and $E(X^2)$ exist . Then

$$= E[X - (\mu_X)^2] = E[x^2 - 2X\mu_X + \mu_X{}^2] = E[x^2] - 2\mu_X E[X] + \mu_X{}^2$$

$$V(X) = E(X^2) - (E(X))^2$$

For example, if X is a continuous random variable with probability density function f_X defined as

$$f_X(x) = \begin{cases} {}^1/_2x, \le x \le 2 \\ 0, \text{ otherwise} \end{cases}$$

then

$$E(X) = \int_{-\infty}^{\infty} xf_X(x)dx = \int_{-\infty}^{2} x({}^1/_2x)dx = {}^4/_3$$

$$E(X^2) = \int_{-\infty}^{\infty} x^2f_X(x)dx = \int_{-\infty}^{2} x^2({}^1/_2x)dx = 2$$

Thus,

$$V(X) = E(X^2) - (E(X))^2 = 2 - ({}^4/_3)^2 = {}^2/_9$$

and the standard deviation,

$$\sqrt{x} = \sqrt{\frac{2}{9}} = \frac{1}{3}\sqrt{2}$$

8.5.4 PROPERTIES OF VARIANCE

1. If X is a continuous random variable defined on a sample space S, and c is any constant, then
   a. $V(cX) = c^2V(X)$.
   b. $V(X + c) = V(X)$.
2. If X and Y are independent continuous random variables defined on the same sample space S, then

$$v(X + Y) = V(X) + V(Y)$$

CHAPTER 9

SPECIAL DISCRETE PROBABILITY MODELS

9.1 BINOMIAL DISTRIBUTION

9.1.1 ORIGIN OF THE DISTRIBUTION

A binomial experiment consists of a fixed number, say n, of Bernoulli trials. If we label the two outcomes as "success" (S) or "failure" (F) with $P(S) = p$ and $P(F) = 1 - p = q$, then in a binomial experiment we are interested in the probability of exactly x successes in n Bernoulli trials.

Assuming independent trials and constant probability of success from trial to trial, each point in the sample space consisting of x S's and $(n-x)$ F's has the probability $p^x q^{n-x}$. Therefore,

$$P(X = x) = \binom{n}{x} p_x q^{n-x}$$

Since a distribution is specified by its parameters, we use the notation $b(x; n, p)$ to designate the binomial distribution with parameters n and p:

$$b(x; n, p) = \binom{n}{x} p^x q^{n-x}$$

$x = 0, 1, 2, \ldots, n.$

For example, $b(x;\ 10,\ 0.3) = \binom{10}{x}(0.3)^x(.07)^{10-x}$

9.1.2 MEAN AND VARIANCE

Let X be a random variable such that X is $b(x; n, p)$, then

1. The mean μ_X or the expected value $E(X)$ of X is given by

$$\mu_X = E(X) = np$$

2. The variance σ^2_X or $V(X)$ of X is given by

$$\sigma^2_X = V(X) = npq$$

9.1.3 APPLICATIONS OF THE BINOMIAL DISTRIBUTION

Before applying the binomial distribution, one must always ensure that the following conditions hold:

1. The random experiment consists of a fixed number, n, of trials.
2. Each trial has only two possible outcomes.
3. The probability of success in each trial is the same for all trials.
4. The trials are independent of each other.

For example, if one third of the voters living in a county are registered as Democrats, then in a sample of 15 voters chosen at random, find the probability that exactly 5 are Democrats, the probability that at least 5 are Democrats, $E(X)$, and $V(X)$, where X is the random variable representing the number of Democrats in n trials.

The selection of a sample of 15 can be considered as a sequence of 15 Bernoulli trials with the selection of a Democrat classified as a success. Since the total voter population is large, the selection can be considered as sampling with replacement. Hence, the trials are independent. The probability of a success = 1/3 on each trial. Therefore,

$$X \text{ is } b(x; 15, 1/3)$$

Thus,

$$P(X = 5) = \left(\frac{15}{5}\right) \times \left(\frac{1}{3}\right)^5 \times \left(\frac{2}{3}\right)^{10}$$

$$= 0.2143$$

If A is the event that at least 5 Democrats are chosen; that is $A = \{5, 6, ..., 15\}$, then $A' = \{0, 1, 2, 3, 4\}$.

$$\begin{aligned}
P(A) &= 1 - P(A') \\
&= 1 - \sum_{x=0}^{4} \binom{15}{x} \left(\frac{1}{3}\right)^{x} \left(\frac{2}{3}\right)^{15-x} = 1 - 0.4041 \\
&= 0.5959 \\
E(X) &= np = 15(1/3) = 5 \\
V(X) npq &= 15(1/3)(2/3) = 10/3
\end{aligned}$$

9.2 WAITING TIME DISTRIBUTION

9.2.1 WAITING FOR THE FIRST SUCCESS

Consider a sequence of Bernoulli trials, and define a random variable Y to be the trial number of the first success (S). Then the number of trials is a random variable, and the sample space does not have a finite number of points in it. Actually,

$$\text{Sample Space} = \{S, FS, FFS, \ldots\}$$

Since $P(S) = p$, and $P(F) = q$, $(q = 1 - p)$, for all trials, the probabilities associated with the outcomes are

$$p, qp, q^2p, q^3p, \ldots, q^np$$

Thus, the probability density function $g(y; p)$ of Y is

$$g(y; p) = q^{y-1}p,$$

$$y = 1, 2, 3, \ldots; 0 \le p \le 1; q = 1 - p$$

This probability density function is called a geometric distribution with parameter p. Note that $g(y; p)$ means the random variable Y has a geometric distribution with parameter p.

> Geometric Distribution
>
> The discrete random variable Y as described above is said to have the geometric distribution with parameter p, $0 \le p \le 1$, if its probability density function $g(y; p)$ satisfies
>
> $$g(y; p) = \begin{cases} p(1 - n)^{y-1}, & y = 1, 2, 3, \ldots \\ 0 & \text{otherwise} \end{cases}$$

Note that

$$\sum_{y=1}^{\infty} g(y;\,p) = p\sum_{y=1}^{\infty} (1-p)^{y-1} = \frac{p}{1-(1-p)} = 1$$

and

$$P_Y(Y = y) = 1 - q^y$$

The tail probability, $P(Y > c)$, can be expressed in a form depending only on c and p

> If Y is $g(y;\,p)$, then
>
> $$P_Y(Y > c) = (1 - p)^c$$
>
> for $c = 0, 1, 2, 3, \ldots$

For example, suppose that a computer is repeatedly printing a long stream of binary digits in which 0 and 1 are equiprobable. The number of digits after each 0 until the next zero is encountered must have a geometric distribution with parameter 1/2. The probability of waiting time of y digits is $\frac{1}{2^y}$, for $y = 1, 2, 3, \ldots$

9.2.2 MEAN AND VARIANCE

> If Y is $g(y;\,p)$, then
>
> a. The mean μ_Y or expected value $E(Y)$ of Y is given by
>
> $$\mu_Y = E(Y) = 1/p$$
>
> b. The variance σ^2_y or $V(Y)$ of Y is given by
>
> $$\sigma^2_y = V(Y) = \frac{g}{p^2}$$

For example, suppose that the relative frequency of a particular blood type is 0.01. A search is made with this particular blood type by repeated sampling with replacement. What is the expected number of trials necessary to obtain a blood specimen of this type and what is the probability that more than 150 typings are needed before finding the correct type?

If we let Y be the trial number on which the correct blood type appears. Then Y is $g(y;\,0.01)$, $E(Y) = 1/(0.01) = 100$, and

$$P_Y(Y > 150) = (1 - 0.01)^{150}$$

$$= 0.22$$

9.2.3 WAITING FOR K SUCCESSES (KS'S)

Consider a sequence of Bernoulli trials in which we wait for k sucesses. In order to have k successes, we must have at least k trials. Hence, the random variable X has the value set

$$\{k, k + 1, k + 2, \ldots\},$$

and every point in the sample space $S = \{k, k + 1, k + 2, \ldots\}$ consists of k S's and $(x - k)$ F's. Hence, the probability of each outcome is

$$p^k(1 - p)^{x-k}$$

That is, the probability density function of X is

$$nb(x; p, k) = \binom{x-1}{k-1} p^k \times (1-p)^{x-k}$$

where $x = k, k + 1, k + 2, \ldots$. This probability density function is said to have a negative binomial distribution with parameters p and k.

> Negative Binomial Distribution
>
> The discrete random variable described above is said to have the negative binomial distribution with parameters p and k where k is a positive integer and $0 \le p \le 1$, if its probability density function satisfies;
>
> $$nb(x; p, k) = \binom{x-1}{k-1} p^k \times (1-p)^{x-k}$$
>
> $x = k, k + 1, k + 2, \ldots$

The mean μ_X or expected value $E(X)$, and the variance σ^2_X or $V(X)$ of this distribution are given as follows:

> If X is $nb(x; p, k)$, then
>
> a. $\mu_X = E(X) = k/p$
>
> b. $\sigma^2_X = V(X) = \dfrac{k(1-p)}{p^2}$

For example, a balanced coin is tossed repeatedly until 3 heads are obtained. What is the probability that the third head is obtained before the sixth trial?

Let X be the trial number, then X is $nb(x;\ 1/2,\ 3)$.

$$P_X(X < 6) = \sum_{x=3}^{5} nb(x;1/2,3) = \frac{1}{2^3} + \frac{3}{2^4} + \frac{6}{2^5} = 12$$

and

$$E(X) = k / p = \frac{3}{12} = 6, \text{ and } V(X) = \frac{3(1-1/2)}{(1/2)^2} = 6$$

9.3 POISSON DISTRIBUTION

> The discrete random variable X is said to have the Poisson distribution with positive parameter λ, if its probability density function satisfies:
>
> $$P(x;\lambda) = \begin{cases} \dfrac{e^{-\lambda}\lambda^x}{x!}, & x = 0,\ 1,\ 2,\ 3,\ldots \\ 0, \text{ otherwise} \end{cases}$$

Recall that $e^{\lambda} = 1 + \frac{\lambda}{1!} + \frac{\lambda^2}{2!} + \frac{\lambda^3}{3!} + \ldots = \sum_{x=0}^{\infty} \frac{\lambda^x}{x!}$

Hence, $\sum_{x=0}^{\infty} \frac{\lambda^x}{x!} e^{-\lambda} = e^{-\lambda} \sum_{x=0}^{\infty} \frac{\lambda^x}{x!} = e^{-\lambda} e^{\lambda} = 1$

The mean μ_X or $E(X)$ and the variance $\sqrt{X}^2$ or $V(X)$ of the random variable X are given as follows:

> If X is $P(x;\ \lambda)$, then
>
> a. $\mu_X = E(X) = \lambda$.
>
> b. $\sigma_X{}^2 = V(X) = \lambda$.

Suppose the chance of triplets in human births is 0.0001. What is the probability of observing at least four sets of triplets in 10,000 births ?

Here n = 10,000, and p = 0.0001. We need the probability $P(X \geq 4)$.

$$P_X(X \geq 4) = \sum_{x=4}^{10,000} \binom{10,000}{x} (0.0001)^x (1-0.0001)^{10,000-x}$$

Note that the random variable X is $b(x;\ 10,000 > 0.0001)$ and determination of probabilities such as $P(X \geq 4)$ involves the evaluation of the binomial coefficients

$$\binom{10{,}000}{x}, \text{ where } x = 4,\ 5,\ 6, \ldots,\ 10{,}000$$

and powers of $p = 0.0001$ for corresponding values of X. Thus, it is difficult to evaluate such probabilities. However, as n becomes very large ($n \to \infty$) with p very small ($p \to 0$) in such a way that p is proportional to $\frac{1}{n}$, the binomial distribution can be approximated by the Poisson distribution.

Thus, in this example, using the Poisson approximation with $\lambda = np = 1$, we have

$$P_X(X \geq 4) = \sum_{x=4}^{\infty} e^{-1} \times \frac{(1)^x}{x!} = 1 - \sum_{x=0}^{3} e^{-1} \times \frac{(1)^x}{x!} = 1 - 0.981 = 0.019$$

9.4 HYPERGEOMETRIC DISTRIBUTION

9.4.1 ORIGIN OF THE DISTRIBUTION

Consider experiments which are made up of a fixed number of trials in which there are two possible outcomes for each trial. For example, consider a box with N balls of which m are red and $(n - m)$ are blue. If we take a sample of n balls from this box without replacement, then

1. The probability of a red ball is not m/N for all trials.
2. The draws are not independent.

For example, a population of N people contains an equal number of men and women. Suppose a sample of size 3 is drawn, what is the probability of X, the number of women in the sample? Find $E(X)$, and $V(X)$ if $N = 20$.

Since the number of men equals to the number of women, it follows that

$$h(x; \frac{N}{2}, N, 3) = \frac{\binom{N/2}{x}\binom{N/2}{3-x}}{\binom{N}{3}}$$

Note that

$$P_X(X = 0) = P_X(X = 3), \text{ and } P_X(X = 1) = P_X(X = 2)$$

Thus, if $N = 20$, then

$$h(0;10,20,3) = \frac{\binom{10}{0} \times \binom{10}{3}}{\binom{20}{3}} = 0.105 = h(3;10,20,3)$$

and

$$h(1;\ 10,\ 20,\ 3) = h(2;\ 10,\ 20,\ 3) = 0.395, \text{ and so forth .}$$

Also, if $N = 20$, then

$$E(X) = n \times \left(\frac{m}{N}\right) = 3\left(\frac{10}{20}\right) = 15$$

and

$$V(X) = \frac{3}{2}\left(1 - \frac{10}{20}\right)\left(\frac{20-3}{20-1}\right) = 0.671$$

9.5 SUMS OF BINOMIAL RANDOM VARIABLES

> Let X_1, and X_2 be two independent binomial random variables such that x_1 is $b(x_1;\ n_1, p)$, X_2 is $b(x_2;\ n_2, p)$, and let $T = X_1 + X_2$. Then the value set of T is 0, 1,..., $n_1 + n_2$. If $T = k$, when $x_1 = i$ and $x_2 = k - i$ for $i = 0, 1, 2, 3,\ldots, k$, then,
>
> $$P_T(T = k) = p^k \times q^{(n_1+n_2)-k} \times \binom{n_1+n_2}{k}, q = 1 - p,$$
>
> or T is $b(t;\ n_1 + n_2, p)$.

This result can be extended to r independent series of trials, each based on the same probability of success.

> Let X_i be $b(x_i;\ n_i, p)$ for $i = 1, 2, 3,\ldots, r$ with X_j and X_k independent for all j and k.

$$\text{If } T = \sum_{i=1}^{r} X_i \text{ then, } T \text{ is } b(t; \sum_{i=1}^{r} n_i, p)$$

For example, suppose that Jim is shooting at a target with a probability of 0.3 that he hits the target. Let X_1 and X_2 count the number of times he is successful in three trials on Monday and four trials on Tuesday respectively. Then X_1 is $b(x_1; 3, 0.3)$ and X_2 is $b(x_2; 4, 0.3)$. If we are interested in the number of times Jim hits the target in the seven trials, then $T = X_1 + X_2$, and T is $b(t; 7, 0.3)$.

Now, for large n and small p, we have seen that the binomial distribution can be approximated by the Poisson distribution with parameter $\lambda = np$. Suppose that Jim is a novice with $p = 0.01$. In order to improve his performance, he practices 100 times on Monday and 200 times on Tuesday. Let the symbol "≈" denote approximately, then

$$X_1 \text{ is } b(x_1; 100, 0.01) \approx P(x_1; 1)$$

$$X_2 \text{ is } b(x_2; 200, 0.01) \approx P(x_2; 2), \text{ and}$$

$$T = X_1 + X_2 \text{ is } b(t; 300, 0.0l) \approx P(t; 3)$$

Thus,

If X_1 and X_2 are independent Poisson random variables with parameters 1, and 2, respectively, then

$$P(X_, + X_2 = n) = \sum_{k=0}^{n} P[X_1 = k, X_2 = n-k]$$

$$= \sum_{k=0}^{n} P[X_1 = k] \times P[X_2 = n-k]$$

$$= \sum_{k=0}^{n} e^{-\lambda} \frac{\lambda_1^k}{k!} \times \bar{e}^{\lambda_2} \frac{\lambda_2^{n-k}}{(n-k)!}$$

$$\frac{\bar{e}^{(\lambda_, + \lambda_2)}}{n!} (\lambda_1 + \lambda_2)^n$$

which is a Poisson Distribution with parameter $\lambda_1 + \lambda_2$.

CHAPTER 10

SPECIAL CONTINUOUS PROBABILITY DISTRIBUTIONS

10.1 UNIFORM PROBABILITY DISTRIBUTION UNIFORM PROBABILITY DENSITY FUNCTION

If X is a continuous random variable with probability density function f_X specified by

$$f_X(x) = \begin{cases} \dfrac{1}{b-a}, & a < x < b \\ 0, & \text{otherwise,} \end{cases}$$

then we say that the

1. random variable X has the uniform probability distribution over the interval $[a, b]$, or X has the rectangular distribution with parameters a and b, where $a < b$.
2. function f_X is the uniform probability density function for the interval $[a, b]$.

With the restriction $a < b$, we have $f_X(x) > 0$, and $\int_{\alpha}^{b} f_X(x)dx = 1$.

Also,

If $a < x < b$, then

$$P_X(X \le x) = \int_{\alpha}^{x} f_x(x)dx = \frac{x-a}{b-a}$$

The uniform probability distribution is used to describe experiments in which a point is chosen at random in a segment of length $b - a = \lambda$. Where the probability that the point falls in any sub-interval of length λ_1 is $\frac{\lambda_1}{\lambda}$..

The probability distribution function for the uniform probability distribution is given by

$$F_X(x) = \begin{cases} 0 & , x \leq a \\ \dfrac{x-a}{b-a} & , a < x < b \\ 1 & , b \leq x \end{cases}$$

For example, to find that the probability of a point selected at random on the line segment [-2, 2] lies between 1 and 2, we let X be the uniform random variable n [-2, 2]. Then the probability density function is given by

$$F_X(x) = \begin{cases} \dfrac{1}{b-a} & , -2 \angle x \angle 2 \\ 0 & , \text{otherwise} \end{cases}$$

$$= \begin{cases} 1/4 & , -2 \angle x \angle 2 \\ 0 & , \text{otherwise} \end{cases}$$

The probability distribution function is given by

$$F_X(x) = \begin{cases} 0 & , x \angle -2 \\ \dfrac{x+2}{4} & , -2 \angle x \angle 2 \\ 1 & , x \trianglerighteq 2 \end{cases}$$

Thus, $P_X(1 \angle X \angle 2) = F_X(2) - F_X(1) = 1 - 3/4 = 1/4$

10.1.2 EXPECTATION AND VARIANCE

Let X be a uniform random variable on the interval $[a, b]$. Then

1. The mean μ_X or the expected value $E(X)$ of X is given by

$$\mu_X = E(X) = \frac{a+b}{2}$$

2. The variance $\sigma_X{}^2$ or $V(X)$ of X is given by

$$\sigma_X^2 = V(X) = \frac{(ba)^2}{12}$$

For example, suppose we choose at random n numbers from the interval [0, 1] with uniform distribution, then if X describes the ith choice, we have

$$\mu_X = E(X) = \frac{0+1}{2} = 1/2$$

and

$$\sqrt{x}^2 = V(X) = \frac{(1-0)^2}{12} = \frac{1}{12}$$

10.2 THE NORMAL DISTRIBUTION

10.2.1 NORMAL PROBABILITY DENSITY FUNCTION

A continuous random variable X is said to have the normal distribution, $-\infty \angle x \angle \infty$, with parameters μ and σ^2, written X is $N(\mu; \sigma^2)$, if its probability density function $f_X(x)$ satisfies

$$f_X(x) = \frac{1}{(\sqrt{2\Pi})\sigma} \times e^{\frac{(x-\mu)^2}{2\sigma^2}}$$

The distribution function $F_X(x)$ is given by

$$F_X(x) = \int_{-\infty}^{x} \frac{1}{(\sqrt{2\Pi})\sigma} \times e^{\frac{(x-\mu)^2}{2\sigma^2}} \times dx$$

The normal distribution defined by the equation

$$f_X(x) = \frac{1}{(\sqrt{2\Pi})\sigma} \times e^{\frac{(x-\mu)^2}{2\sigma^2}}$$

represents a family of distributions with the specific number of that family being determined by the values of the parameters μ and σ^2. Graphically, the normal curve is bell-shaped and symmetrical around $x = \mu$. The following diagram shows three normal probability distributions.

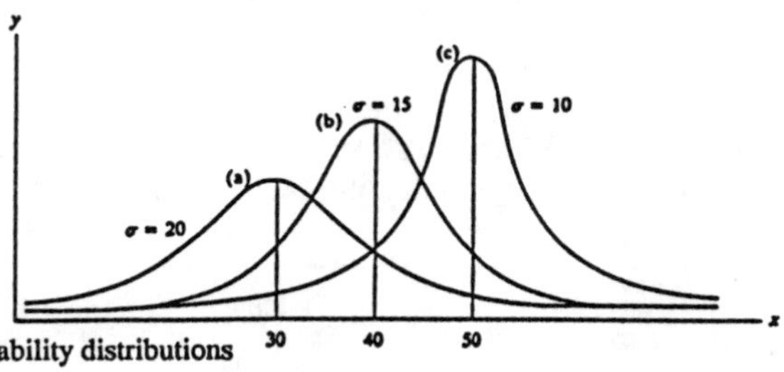

Three normal probability distributions

10.2.2 EXPECTATION AND VARIANCE

Let X be a normal random variable with parameters μ and σ^2. That is, X is $N(\mu; \sigma^2)$, then

1. The expected value of X is given by

$$E(X) = \mu$$

2. The variance of X is given by

$$V(X) = \sigma^2$$

10.2.3 STANDARDIZED NORMAL DISTRIBUTION

In the normal probability distribution, the $P_X(X \angle x)$ is of course the area under the curve of $F_X(x)$ from minus infinity to x. That is

$$P_X(X \angle x) = \int_{-\infty}^{x} f_X(x)dx$$

If $Z = \dfrac{X-\mu}{\sigma}$, then

$$F_Z(Z) = P_Z(Z \angle z) = P_X\left(\frac{X-\mu}{\sigma} \angle Z\right) = P_X(X \angle \sigma_Z + \mu) =$$

$$\int_{-\infty}^{\sigma_Z+\mu} \frac{1}{\sqrt{2\Pi}\sigma} e^{-\frac{1}{2}\left(\frac{x-\mu}{5}\right)^2} dx = \int_{-\infty}^{Z} \frac{1}{\sqrt{2\Pi}} e^{z^2/2} dz$$

Hence, we obtain the standardized normal distribution or curve

$$\phi_Z = \frac{1}{\sqrt{2\Pi}} \times \bar{e}^{\frac{z^2}{2}}$$

Which has mean $\mu = 0$ and variance $\sigma^2 = 1$. That is, the random variable Z is $N(0; 1)$. The function $\varnothing_Z$ is called the probability density function of a normal distribution with parameters μ = 0, and $\sigma^2 = 1$.

Thus if the random variable $Z = \dfrac{X-\mu}{\sigma}$, and $z = \dfrac{X-\mu}{\sigma}$, then the probability distribution function of the standardized normal random variable Z is given by

$$\Phi_Z(Z \leq z) = \int_{-\infty}^{z} \varnothing_Z(z)dz$$

$$= \frac{1}{\sqrt{2\Pi}} \int_{-\infty}^{z} e^{-z^2/2} dz$$

The graph of this distribution appears below

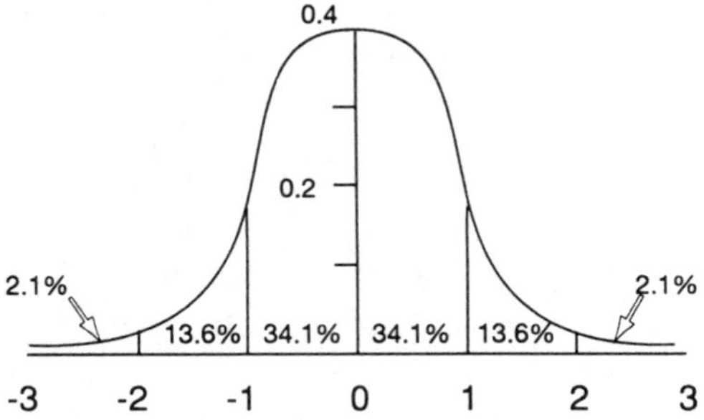

Normal distribution $N(0, 1)$

Tables for both $\varnothing_Z(z)$ and $\Phi_Z(z)$ are given on pages 91 and 92 respectively. Since the curve of $\varnothing_Z(z)$ is symmetrical about $z = 0$, we can obtain the area between any two values of z.

10.2.4 PROBABILITIES OF NORMAL VARIABLES

Let X be a continuous random variable such that X is $N(\mu;\sigma^2)$. We compute the probability that X lies between a and b, denoted by $P_X(a \angle x \angle b)$, as follows.

> $P_X(a \angle X \angle b) = P_Z(a' \leq Z' \leq b') = \Phi_Z(b') - \Phi_Z(a')$, where Z is the standardized random variable corresponding to X and hence Z in $N(0; 1)$.

Note that $a' = a' = \dfrac{b-\mu}{\sigma}$ and $b' = \dfrac{b-\mu}{\sigma}$

For example, suppose that a continuous random variable X is $N(3; 16)$. Calculate $P_X(X \leq 9)$ and $P_X(-6 \leq X \leq 9)$.

If X is $N(3; 16)$, then $Z = \dfrac{X-3}{4}$ is $N(0; 1)$. Thus,

$$P_X(X \le 9) = \Phi_Z\left(Z \angle \frac{9-3}{4}\right)$$
$$= \Phi_Z(Z \angle 1.5) = 0.9332$$

and

$$P_X(-6 \angle X \angle 9) = \Phi_Z\left(\frac{-6-3}{4}\right) \angle Z \angle \frac{9-3}{4}$$
$$= \Phi_Z(-1.5 \le Z \le 1.5)$$
$$= \Phi_Z(Z \le 1.5) - \Phi_Z(Z - 1.5)$$
$$= \Phi_Z(Z \le 1.5) - \Phi_Z(Z \ge -1.5)$$
$$= \Phi_Z(Z \le 1.5) - (-\Phi_Z(Z \le -1.5))$$
$$= 0.9332 - (1 - 0.9332)$$
$$= 0.9332 - 0.0668$$
$$= 0.8664$$

10.2.5 ORDINATES OF NORMAL CURVE

$$\phi(z) = \frac{1}{\sqrt{2\Pi}} e^{-1/2} z^2$$

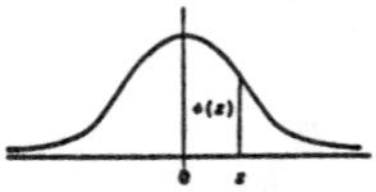

(TO FOUR DECIMAL PLACES)

Z	.00	.01	.02	.03	.04	.05	. 06	.07	.08	.09
.0	.3989	.3989	.3989	.3988	.3986	.3984	.3982	.3980	.3977	.3973
.1	.3970	.3965	.3961	.3956	.3951	.3945	.3939	.3932	.3925	.3918
.2	.3910	.3902	.3894	.3885	.3876	.3867	.3857	.3847	.3836	.3825
.3	.3814	.3802	.3790	.3778	.3765	.3752	.3739	.3725	.3712	.3697
.4	.3683	.3668	.3653	.3637	.3621	.3605	.3589	.3572	.3555	.3538
.5	.3521	.3503	.3485	.3467	.3448	.3429	.3410	.3391	.3372	.3352
.6	.3332	.3312	.3292	.3271	.3251	.3230	.3209	.3187	.3166	.3144
.7	.3123	.3101	.3079	.3056	.3034	.3011	.2989	.2966	.2943	.2920
.8	.2897	.2874	.2850	.2827	.2803	.2780	.2756	.2732	.2709	.2685
.9	.2661	.2637	.2613	.2589	.2565	.2541	.2516	.2492	.2468	.2244
1.0	.2420	.2396	.2371	.2347	.2323	.2299	.2275	.2251	.2227	.2203
1.1	.2179	.2155	.2131	.2107	.2083	.2059	.2036	.2012	.1989	.1965
1.2	.1942	.1919	.1895	.1872	.1849	.1826	.1804	.1781	.1758	.1736
1.3	.1714	.1691	.1669	.1647	.1626	.1604	.1582	.1561	.1539	.1518
1.4	.1497	.1476	.1456	.1435	.1415	.1394	.1374	.1354	.1334	.1315
1.5	.1295	.1276	.1257	.1238	.1219	.1200	.1182	.1163	.1145	.1127
1.6	.1109	.1092	.1074	.1057	.1040	.1023	.1006	.0989	.0973	.0957
1.7	.0940	.0925	.0909	.0893	.0878	.0863	.0848	.0833	.0818	.0804
1.8	.0790	.0775	.0761	.0748	.0734	.0721	.0707	.0694	.0681	.0769
1.9	.0656	.0644	.0632	.0620	.0608	.0596	.0584	.0573	.0562	.0551
2.0	.0540	.0529	.0519	.0508	.0498	.0488	.0478	.0468	.0459	.0449
2.1	.0440	.0431	.0422	.0413	.0404	.0396	.0387	.0397	.0371	.0363
2.2	.0355	.0347	.0339	.0332	.0325	.0317	.0310	.0303	.0297	.0290
2.3	.0283	.0277	.0270	.0264	.0258	.0252	.0246	.0241	.0235	.0229
2.4	.0024	.0219	.0213	.0208	.0203	.0198	.0194	.0189	.0184	.0180
2.5	.0175	.0171	.0167	.0163	.0158	.0154	.0151	.0147	.0143	.0139
2.6	.0136	.0132	.0129	.0126	.0122	.0119	.0116	.0113	.0110	.0107
2.7	.0104	.0101	.0099	.0096	.0093	.0091	.0088	.0086	.0084	.0081
2.8	.0079	.0077	.0075	.0073	.0071	.0069	.0067	.0065	.0063	.0061
2.9	.0060	.0058	.0056	.0055	.0053	.0051	.0050	.0048	.0047	.0046
3.0	.0044	.0043	.0042	.0040	.0039	.0038	.0037	.0036	.0035	.0034
3.1	.0033	.0032	.0031	.0030	.0029	.0028	.0027	.0026	.0025	.0025
3.2	.0024	.0023	.0022	.0022	.0021	.0020	.0020	.0019	.0018	.0018
3.3	.0017	.0017	.0016	.0016	.0015	.0015	.0014	.0014	.0013	.0013
3.4	.0012	.0012	.0012	.0011	.0011	.0010	.0010	.0010	.0009	.0009
3.5	.0009	.0008	.0008	.0008	.0008	.0007	.0007	.0007	.0007	.0006
3.6	.0006	.0006	.0006	.0005	.0005	.0005	.0005	.0005	.0005	.0004
3.7	.0004	.0004	.0004	.0004	.0004	.0004	.0003	.0003	.0003	.0003
3.8	.0003	.0003	.0003	.0003	.0003	.0002	.0002	.0002	.0002	.0002
3.9	.0002	.0002	.0002	.0002	.0002	.0002	.0002	.0002	.0001	.0001
4.0	.0001	.0001	.0001	.0001	.0001	.0001	.0001	.0001	.0001	.0001
z	.00	.01	.02	.03	.04	.05	.06	.07	.08	.09

10.2.6 STANDARD NORMAL DEVIATION FUNCTION

$$\Phi(z) = \int_{-\infty}^{-z} \frac{1}{\sqrt{2\Pi}} e^{-Z^2/Z} = P(Z \le z)$$

z	0	1	2	3	4	5	6	7	8	9
0.0	0.5000	0.5040	0.5080	0.5120	0.5160	0.5199	0.5239	0.5279	0.5319	0.5359
0.1	0.5398	0.5438	0.5478	0.5517	0.5557	0.5596	0.5636	0.5675	0.5714	0.5753
0.2	0.5793	0.5832	0.5871	0.5910	0.5948	0.5987	0.6026	0.6064	0.6103	0.6141
0.3	0.6719	0.6217	0.6255	0.6293	0.6331	0.6368	0.6406	0.6443	0.6480	0.6517
0.4	0.6554	0.6591	0.6628	0.6664	0.6700	0.6736	0.6772	0.6808	0.6844	0.6879
0.5	0.6915	0.6950	0.6985	0.7019	0.7054	0.7088	0.7123	0.7157	0.7190	0.7224
0.6	0.7257	0.7291	0.7324	0.7357	0.7389	0.7422	0.7454	0.7486	0.7517	0.7549
0.7	0.7580	0.7611	0.7642	0.7673	0.7703	0.7734	0.7764	0.7794	0.7823	0.7852
0.8	0.7881	0.7910	0.7939	0.7967	0.7995	0.8023	0.8051	0.8078	0.8106	0.8133
0.9	0.8159	0.8186	0.8212	0.8238	0.8264	0.8289	0.8315	0.8340	0.8365	0.8389
1.0	0.8413	0.8438	0.8461	0.8485	0.8508	0.8531	0.8554	0.8577	0.8599	0.8621
1.1	0.8643	0.8665	0.8686	0.8708	0.8729	0.8749	0.8770	0.8790	0.8810	0.8830
1.2	0.8849	0.8869	0.8888	0.8907	0.8925	0.8944	0.8962	0.8980	0.8997	0.9015
1.3	0.9032	0.9049	0.9066	0.9082	0.9099	0.9115	0.9131	0.9147	0.9162	0.9177
1.4	0.9192	0.9207	0.9222	0.9236	0.9251	0.9265	0.9278	0.9292	0.9306	0.9319
1.5	0.9332	0.9345	0.9357	0.9370	0.9382	0.9394	0.9406	0.9418	0.9430	0.9441
1.6	0.9452	0.9463	0.9474	0.9484	0.9495	0.9505	0.9515	0.9525	0.9535	0.9545
1.7	0.9554	0.9564	0.9573	0.9582	0.9591	0.9599	0.9608	0.9616	0.9625	0.9633
1.8	0.9641	0.9648	0.9656	0.9664	0.9671	0.9678	0.9686	0.9693	0.9700	0.9706
1.9	0.9713	0.9719	0.9726	0.9732	0.9738	0.9744	0.9750	0.9756	0.9762	0.9767
2.0	0.9772	0.9778	0.9783	0.9788	0.9793	0.9798	0.9803	0.9808	0.9812	0.9817
2.1	0.9821	0.9826	0.9830	0.9834	0.9838	0.9842	0.9846	0.9850	0.9854	0.9857
2.2	0.9861	0.9864	0.9868	0.9871	0.9874	0.9878	0.9881	0.9884	0.9887	0.9890
2.3	0.9893	0.9896	0.9898	0.9901	0.9904	0.0906	0.9909	0.9911	0.9913	0.9916
2.4	0.9918	0.9920	0.9922	0.9925	0.9927	0.9929	0.9931	0.9932	0.9934	0.9936
2.5	0.9938	0.9940	0.9941	0.9943	0.9945	0.9946	0.9948	0.9949	0.9951	0.9952
2.6	0.9953	0.9955	0.9956	0.9957	0.9959	0.9960	0.9961	0.9962	0.9963	0.9964
2.7	0.9965	0.9966	0.9967	0.9968	0.9969	0.9970	0.9971	0.9972	0.9973	0.9974
2.8	0.9974	0.9975	0.9976	0.9977	0.9977	0.9978	0.9979	0.9979	0.9980	0.9981
2.9	0.9981	0.9982	0.9982	0.9983	0.9984	0.9984	0.9985	0.9985	0.9986	0.9986
3.0	0.9987	0.9990	0.9993	0.9995	0.9997	0.9998	0.9998	0.9999	0.9999	1.0000

10.3 EXPONENTIAL PROBABILITY DISTRIBUTION

10.3.1 EXPONENTIAL PROBABILITY FUNCTION

A continuous random variable X is said to have the exponential distribution with positive parameter λ, if its probability density function $f_X(x)$ satisfies:

$$f_X(x) = \begin{cases} \lambda e^{-\lambda X}, & 0 \leq_x \angle\infty \\ 0, & \text{otherwise} \end{cases}$$

Note that we have $f_X(x) > 0$ and $\int_0^\infty f_X(x)dx = \int_0^\infty \lambda e^{-\lambda x}dx$

$$= [-e^{-\lambda x}]^\infty_0$$

$$= 1$$

Further,

$$P_X(0 \leq X \leq x) = \int_0^x \lambda \varepsilon^{-\lambda\xi}dx = 1 - e^{-\lambda\xi}$$

From this, it follows immediately that

$$P_X(0 \leq x \leq X) = 1 - P_X(0 < x < X) = e^{-\lambda\xi}$$

and when $0 \angle x_1 \angle x_2$, we have

$$P_X(x_1 \leq X \leq x_2) = \left(1 - e^{-\lambda x_2}\right) - \left(1 - e^{-\lambda x_1}\right)$$

10.3.2 EXPECTATION AND VARIANCE

Let x be a continuous exponential random variable with parameter λ. Then the

1. Expected value of X is given by $E(X) = \frac{1}{\lambda}$.

2. Variance of X is given by $V(X) = \frac{1}{\lambda^2}$.

For example, for bulbs of a certain type, the time to failure is thought to have an exponential distribution with parameter λ. A number of the bulbs are put on test and it is found that 5% have failed by 100 hours.

Estimate

a. The probability that this type will last more than 200 hours.

b. The probability that one of the survivors of the test will complete 200 hours.

3. $E(X)$ and $V(X)$.

Since an exponential distribution is determined by one parameter, we must estimate λ. This estimate must be based on the 5% which failed before 100 hours. Thus,

a. $P_X(X - x) = 1 - e^{-lx}$. If we estimate $P_X(X - 100)$ as 0.05, we have $1 - e^{-100x} = 0.05$. Thus,

$$100l = 0.051, \text{ or } l = 0.00051$$

b. $P_X(X \geq x) = 1 - P_X(X \angle x) = e^{-\lambda\xi}$. Hence,

$$P_X(X - 200) = e^{-200\lambda} = (e^{-100\lambda})^2 = (0.95)^2\ (0.9025)$$

where the probability that a new bulb lasts 200 hours equals to

$$1 - 0.05 = 0.95$$

c. $E(X) = \dfrac{1}{\lambda} = \dfrac{1}{0.00051} = 1961$ hours,

and $V(X) = \dfrac{1}{\lambda^2} = 3844675$ hours.

10.4 LAW OF LARGE NUMBERS AND THE CENTRAL LIMIT THEOREM

10.4.1 LAW OF LARGE NUMBERS

Chebyshev's inequality of section 6.6.1 can be extended to continuous random variables.

> Let X be a continuous random variable with expected value $E(X)$ and a variance $V(X)$. Let ε be any positive real number. Then the probability that a value of X occurs that differs from $E(X)$ by more than is less than or equal to $(V(X))/(\varepsilon^2)$, that is,
>
> $$P_X\left(|X - E(X)| \geq \varepsilon\right) \leq \frac{V(X)}{\varepsilon^2}$$

In addition, the law of large numbers for discrete random variables of section 6.6.2 can also be extended to continuous random variables as follows.

Law Of Large Numbers

Let X be a continuous random variable with mean $E(X)$ and variance $V(X)$. If $\tilde{X}$ is the sample mean for a set of sample random variables of size n for X, then for any $\varepsilon > 0$,

$$\underset{n \to \infty}{\text{limit}}\, P_{\tilde{X}}\left(\left|\tilde{X} - E(X)\right| > \varepsilon\right) = 0$$

10.4.2 NORMAL APPROXIMATION TO THE BINOMIAL DISTRIBUTION

The binomial distribution approaches the normal distribution as n increases. Thus, if the probability of a success in each trial p is not close to 0 or 1, then the binomial distribution can be closely approximated by a normal distribution with the same mean and variance as the binomial distribution; that is, $\mu = np$, and $\sigma = \sqrt{npq}$. In general,

If X is a binomial random variable such that X is $b(x; n, p)$, then for any real numbers a and b,

$$P_X(a \le X \le b) \approx \Phi_Z\left(\frac{b - np + 1/2}{\sqrt{npq}}\right) - \Phi_Z\left(\frac{-np - 1/2}{\sqrt{npq}}\right); n \to \infty$$

where Φ_Z is the normal probability distribution function with mean 0 and variance 1, $\approx$ indicates "approximately equal to," and n is the number of Bernoulli trials.

For example, suppose that a fair die is tossed 5000 times. The probability that the number of tosses that result in a 6 between 850 and 900 is the sum

$$\sum_{k=850}^{900} \binom{5000}{k} \left(\frac{1}{6}\right)^k \left(\frac{5}{6}\right)^{5000-k}$$

An approximation to this sum is provided by the normal approximation to the binomial where $a = 850$, $b = 900$, $p = 1/6$, and $n = 5000$. Thus,

$$P_X(850 < x < 900) \approx \Phi_Z\left(\frac{b - np - 1/2}{\sqrt{npq}}\right) - \Phi_Z\left(\frac{a - np - 1/2}{\sqrt{npq}}\right)$$
$$= \Phi_Z(2.55) - \Phi_Z(0.61)$$

From the table of the standard normal distribution, (see page 92), we obtain,

$$P_X(850 < X < 900) = (2.55) - F_Z(0.61)$$
$$= 0.9946 - 0.7291$$
$$= 0.2655$$

10.4.3 CENTRAL LIMIT THEOREM

The normal approximation of the binomial distribution can be made for a sample mean of a general random variable X. One can assert that the sample mean X is approximately a normal random variable. This can be stated in the following important theorem in probability theory.

Central Limit Theorem

If $\tilde{X}_n$ is the sample mean of a sample of size n for a random variable X with mean, μ and variance σ^2, then for any real numbers a and b with $a < b$,

$$\underset{n \to \infty}{\text{limit}}\, P_{\tilde{X}}\left(a < \frac{\tilde{X}_n - \mu}{\frac{\sigma}{\sqrt{n}}} < b \right) = \frac{1}{2\Pi} \int_a^b e^{-Z^2/2} dz$$

For example, the time a certain freshman arrives at his 8:00 o'clock class seems to be uniform over the interval 7:58 AM and 8:04 AM. A sample of size 100 of the arrival times is taken. Find the approximate probability that the sample mean will lie between 7:58 and 8:01.

Let X be the uniform random variable over the interval [0, 6], where 0 corresponds to the time 7:58 AM. Then

$$E(X) = 3, \text{ and } V(X) = 3$$

Thus, the event $1 < \tilde{X}_n < 3$ is equivalent to the event

$$\frac{1-3}{\frac{\sqrt{3}}{10}} < \frac{\tilde{X}_n - 3}{\frac{\sqrt{3}}{10}} < \frac{3-3}{\frac{\sqrt{3}}{10}}$$

Thus, with a = -11.5, and b = 0, using the central limit theorem, we obtain

$$\begin{aligned} P_{\tilde{X}_n}\left(1 \langle \tilde{X}_n \langle 3\right) &= \Phi_Z(0) - \Phi_Z(-11.5) \\ &= 0.05 - 0.0 \\ &= 0.50 \end{aligned}$$

Note that $\Phi_Z(-11.5)$ is positive, but so small that it is negligible; that is, $\Phi_Z(-11.5) \approx 0$.

CHAPTER 11

GENERATING FUNCTIONS

11.1 GENERATING FUNCTIONS FOR DISCRETE DENSITIES

11.1.1 MOMENTS

If X is a random variable with a finite range $\{x_1, x_2, \ldots, x_n\}$ and a probability density function $f = f_X$, and if its mean $\mu = E(X)$ and its variance $\sigma^2 = V(X)$, then f can be determined completely in terms of the moments of X, which are numbers defined as follows.

> Moments
>
> For each positive integer k, the kth moment of X, denoted by u_k is defined by
>
> $$u_k = E(X^k)$$

Here, $f(x_i) = P(X = x_i)$. The first moment of X, u_1, is just the mean of X. That is $u_1 = \mu$.

11.1.2 CENTRAL MOMENTS

We have seen that a function $Y = g(x)$ of a random variable X is itself a random variable and admits a probability distribution of its own related to that of X. In terms of moments of the random variable X, if $g(x) = (x - \mu)^k$, then we can define the central moment of X as follows.

> Central Moment
>
> For each positive integer k, the kth central moment of X, denoted by u_k, is defined by
>
> $$u_k = E((X - \mu)^k)$$

For $k = 1$, $\mu_1 = E(X - \mu)^1 = E(X) - \mu = \mu - \mu = 0.$

The second central moment of È, μ_2 is of special interest. It is the variance of X and is typically denoted by σ^2 or σ_X^2 or sometimes $V(X)$. That is $\sigma^2 = \mu^2 - \mu_1{}^2$.

11.1.3 n^th FACTORIAL MOMENTS

When factorials are involved in a probability distribution, such as the binomial and Poisson, it is often convenient to calculate $E(X(X - 1))$, and then use the fact that $X(X - l) = X^2 - X$ and the linearity of E for finding $E(X)$, and, ultimately, σ^2. The reason is that $x(x - l)$ divides $x!$ to leave $x(x - 2)!$ which can be very convenient. This moment is a special case of a general type of moment called n^{th} factorial moment.

> n^{th} Factorial Moment
>
> The nth factorial moment of a random variable X is, for each positive integer n,
>
> $$E(X(X - 1)(X - 2) \ldots (X - n + 2))$$

For example, if X is Poisson with parameter λ, then

$$E\big(X(X-1)\big) = \sum_{x=0}^{\infty} x(x-1) \times e^{-\lambda} \frac{\lambda^X}{x!} = e^{-\lambda} \sum_{x=2}^{\infty} \frac{\lambda^2}{(x-2)!},$$

since the terms for $x = 0$, and $x = 1$ vanish.

Now, if we let $y = x - 2$, so $x = y + 2$, and

$$E\big(X(X-1)\big) \approx e^{-\lambda} \sum_{y=0}^{\infty} \frac{\lambda^{y+2}}{y!} = \lambda^2 \sum_{y=0}^{\infty} e^{-\lambda} \frac{\lambda^y}{y!} = \lambda^2, y$$

then since

$$\lambda^2 = E(X(X - l)) = E(X^2) - E(X) = E(X^2) - \lambda,$$

it follows that

$$E(X^2) = \lambda^2 + \lambda.$$

and, hence

$$\sigma^2 = \lambda^2 + \lambda - \mu_1^2 = \lambda.$$

Recall that for Poisson distribution the mean and variance are the same.

11.1.4 MOMENT–GENERATING FUNCTIONS

Moment-generating functions provide a method of characterizing a probability distribution. When this method can be used, it provides one of the most powerful tools a probabilist has at his or her disposal. As a by-product, moment-generating functions are sometimes useful for computing means, standard deviations, and other characteristics of distributions.

> Moment-Generating Functions
>
> If X is a random variable, then its moment generating function, $M_X(t)$, is defined as
>
> $$M_X(t) = E(e^{Xt})$$
>
> provided that this expectation exists for some interval of t which includes the origin. That is, there is a positive number c, so that whenever $-c < t < c$, $E(e^{Xt})$ is finite.

When M exists as required in the above definition, all the derivatives of M_X exist at the origin, and the coefficients of $\frac{t}{k!}$ in the series expansion of M_X is the kth derivative $M(k)$ of M_X For example, if a random variable X is $b(x; n, p)$ for which

$$f_X(x) = \binom{n}{x} p^x q^{n-x}, \quad x = 0,\ 1,\ 2, \ldots,\ \text{n}$$

then

$$\begin{aligned} M_X(t) &= E\left(e^{Xt}\right) = \sum_x e^{xt} f_X(x) \\ &= \sum_{x=0}^{n} e^{xt} \binom{n}{x} p^x q^{n-x} \\ &= \sum_{x=0}^{n} \binom{n}{x} \left(e^t p\right)^x q^{n-x} = \left(q + pe^t\right)^n \end{aligned}$$

11.1.5 MOMENT THEOREMS

Using the moment-generating function, one can show, at least in the case of a discrete random variable with finite range, that its probability density function is completely determined by its moment-generating function, which is in turn completely determined by its moments.

Theorems:

1. Let X be a discrete random variable with finite range $\{x_1, x_2, \ldots, x_n\}$, and moments $u_k = E(X^k)$. Then the moment series

$$M_X(t) = \sum_{k=0}^{\infty} \frac{u_k t^k}{k!}$$

converges for all t to an infinitely differentiable function $MX(t)$.

2. Let X be a discrete random variable with finite range $\{x_1, x_2, \ldots, x_n\}$, probability density function f_X and moment-generating function M_X. Then M_X is uniquely determined by f_X and conversely.

3. Let X and Y be random variables having moment-generating functions M_X and M_y respectively. Then X and Y are identically distributed (have the same probability density) if and only if

$$M_X(t) = M_y(t)$$

for all t in some neighborhood of the origin 0.

11.1.6 ORDINARY-GENERATING FUNCTIONS

Let X be a random variable with a finite range $x_1, x_2, \ldots, x_n$, a probability distribution function f_X and a moment-generating function $M_X(t)$. In the special but important case where the x_i's are all nonnegative integers; that is, $x_i = i$, we have

$$M_X(t) = e^{ti} f_X(i)$$

and $M_X(t)$ is a polynomial in e^t.

If we write $z = e^t$, and define the function h by

$$h(z) = \sum_{i=0}^{n} z^i f_X(i),$$

then $h(z)$ is a polynomial in z containing the same information as $MX(t)$ and in fact

$$h(z) = M_X(\log z)$$

$$M_X(t) = h(e^t)$$

The function $h(z)$ is called the ordinary-generating function for X.

Note that $h(1) = M_X(0) = 1$, $h'(1) = M_X'(0) = u_1$, and $h''(1) = M_X''(0) = u_2 - u_1$. It follows from all this that if we know *MX(t)*, then we know *h(z)*, and if we know *h(z)*, then we can find $f_X(i)$ by Taylor's formula :

$$f_X(i) = \text{Coefficient of } z^i \text{ in } h(z)$$
$$= \frac{h^{(i)}(0)}{i!}$$

For example, suppose that the moments of a certain variable *X* are given by:

$u_0 = 1,\ u_k = \frac{1}{2} + \frac{2^k}{4}$, for $k \geq 1$. Then,

$$M_X(t) = \sum_{k=0}^{\infty} \frac{u_k t^k}{k!} = 1 + \frac{1}{2}\sum_{k=1}^{\infty} \frac{t^k}{k!} + \frac{1}{4}\sum_{k=1}^{\infty} \frac{(2t)^k}{k!}$$

$$= \frac{1}{2} + \frac{1}{2}e^t + \frac{1}{4}e^{2t}$$

This is a polynomial in $z = e^t$, and

$$h(z) = \frac{1}{4} + \frac{1}{2}z + \frac{1}{4}z^2$$

Thus, *X* must have range {0, 1, 2}, and f_X must have values: 1/4 , 1/2 , 1/4.

11.1.7 PROPERTIES OF MOMENT-GENERATING AND ORDINARY-GENERATING FUNCTIONS

Both the moment-generating function, M_X, and the ordinary-generating function *h* have many properties useful in the study of random variables, of which we consider only a few here.

Properties:

Let *X* be a discrete random variable and $Y = X + a$, where *a* is any constant . Then

1. $M_y(t) = e^{ta} MX(t)$.
2. If $X = bY$, where *b* is any constant, then $M_y(t) = M_y(bt)$
3. If $Z = \frac{X - \mu}{\sigma}$, then $M_Z(t) = e^{\frac{\mu t}{\sigma}} M_X\left(\frac{t}{\sigma}\right)$

4. If X and Y are independent random variables and $Z = X + Y$ is their sum, with probability density functions f_X, f_y, and f_z respectively, then

$$M_X(t) = M_X(t)\, M_y(t)$$

and

$$h_Z(z) = h_X(z)\, h_y(z)$$

11.2 GENERATING FUNCTIONS FOR CONTINUOUS DENSITIES

11.2.1 MOMENTS

If X is a continuous random variable defined on a probability space S, with probability density function f_X then the nth moment, u_n is defined as

$$u_n = E\left(X^n\right) = \int_{-\infty}^{+\infty} x^n f_X(x)dx$$

provided that this integral converges.

Note that just as in the discrete case, $u_0 = 1$, $u_1 = \mu$ and $u_2 - u^2{}_1 = \sigma^2$.

11.2.2 MOMENT-GENERATING FUNCTIONS

Let X be a continuous random variable defined on a probability space S with probability density function f_X. Then the moment-generating function $M_X(t)$ for X is defined by the formula

$$M_X(t) = \sum_{k=0}^{\infty} \frac{u_k t^k}{k!} = \sum_{k=0}^{\infty} \frac{E\left(X^k\right)t^k}{k!} = E\left(e^{tx}\right)$$
$$= \int_{-\infty}^{+\infty} e^{tx} f_X(x)dx$$

provided this series converges.

As in the discrete case, $u_n = M_X^{(n)}0$. For example, let X be a continuous random variable with range [0, 1] and probability density function $f_X(x)$ = 1, for $0 \angle x \angle x$, (uniform density). Then,

$$u_n = \int_o^1 x^n dx = \frac{1}{n+1}$$

and

$$M_X(t) = \sum_{k=0}^{\infty} \quad = \frac{e^t - 1}{t}$$

Here, the series converges for all t. Alternatively, we have

$$M_X(t) = \sum_{k=0}^{+\infty} e^{tx} f_X(x) = \int_0^1 e^{tx} dx$$

$$= \frac{e^t - 1}{t}$$

Using L'Hopital's rule, we obtain

$$u_0 = M_X = \underset{t \to 0}{\text{limit}} \frac{e^t - 1}{t} = 1$$

$$u_1 = M_{X'}(0) = \underset{t \to 0}{\text{limit}} \frac{te^t - e^t + 1}{t^2} = \frac{1}{2}$$

$$u_2 = M_{X''}(0) = \underset{t \to 0}{\text{limit}} \frac{t^3e^t - 2t^2e^t + 2te^t - 2t}{t^4} = \frac{1}{3}$$

$$\sigma^2 = M_{X''}(0) - (M_{X'}(0))^2 = \frac{1}{3} - \frac{1}{4} = \frac{1}{12}$$

11.2.3 MOMENT THEOREMS

In general, the series defining $M_X(t)$ does not converge for all t. But in the special case where X is bounded; that is, the range of X is contained in a finite interval, the series does converge for all t.

Theorems:

1. Let X be a continuous random variable with range contained in the interval $[-c, c]$. Then the series

$$M_X(t) = \sum_{k=0}^{\infty} \frac{u_k t^k}{k!}$$

converges for all t to an infinitely differentiable function $M_X(t)$, and

$$m_X^{(n)}(0) = u_n$$

2. If X is a bounded random variable, then the moment generating function $M_X(t)$ of X determines the probability density function $f_X(x)$ uniquely.

11.2.4 CHARACTERISTIC FUNCTIONS

Let X be a continuous random variable defined on a sample space S, with a probability density function f_X and a moment-generating function M_X. Then we know that

$$M_X(t) = \sum_{k=0}^{\infty} \frac{u_k t^k}{k!} = \int_{-\infty}^{+\infty} e^{tx} f_x(x)dx$$

provided that this series converges.

If we replace t by iu, where u is a real number and $i = \sqrt{-1}$, then the series converges for all u, and we can define a function $k_X(u)$ of X called the characteristic function of X as follows

Characteristic Functions

The function $k_X(u)$ defined by the formula

$$k_X(u) = M_X(iu) = \int_{-\infty}^{+\infty} e^{iux} f_X(x)dx$$

is called the characteristic function of X. This function k_X is defined by the above equation even if the series does not converge.

The defining equation of k_X says that k_X is a Fourier transform of f_X. Of course, the Fourier transform has an inverse given by the formula

$$f_X(x) = \frac{1}{2\Pi} \int_{-\infty}^{+\infty} e^{-iux} k_X(u)du$$

Note that it can be easily seen that the characteristic function k_X, and hence the moment-generating function M_X, determines the probability density function f_X uniquely.

11.2.5 PROPERTIES OF CHARACTERISTIC FUNCTIONS

Let X be a random variable with characteristic function k_X. Then

1. $k_X(0) = 1$, since $k_X(0) = E(e^0)$.
2. $|k_X(u)| \angle 1$ for all u, since if X has a probability density function f_X, then we have

$$|k_X(u)| = |\int_{-\infty}^{+\infty} e^{iux} f_X(x)dx|$$

$$\angle \int_{-\infty}^{+\infty} |e^{iux}| f_X(x)dx = \int_{-\infty}^{+\infty} f_X(x)dx = 1$$

3. If X has a probability density function f_X and f_X is even; that is, $f_X(-x) = f_X(x)$ then $k_X(u)$ is real-valued for all u.
4. If X is a discrete random variable taking on only integer values, then

$$k_X(u + 2\Pi) = k_X(u)$$

CHAPTER 12

MARKOV CHAINS

12.1 INTRODUCTION

12.1.1 STOCHASTIC PROCESSES

Many applications of probability models are used in connection with phenomena whose characteristics evolve in time or change with respect to some index such as volume or distance. For example, it may be desired to model the depth of a river at time t, as measured from some fixed origin. If we are interested in future depths, in an attempt to predict the frequency and duration of floods, it is natural to consider depths at future times, t_1, t_2, t_3, and so on as observed values of random variables X_{t_1}, X_{t_2}, X_{t_3}, and so on.

Since the set of possible future times is large, perhaps the interval $(0, \infty)$, we may be concerned with infinitely many jointly distributed random variables, say, those in the class $\{X_t: t > 0\}$.

Definitions: Stochastic Process

A set $\{X_t, t \in T\}$ of random variables is called a stochastic process. The set T is called the index set of the stochastic process. If the stochastic set is finite, say, $T = \{t_1, t_2, t_3, \ldots, t_n\}$, then the stochastic process $\{X_t: t \in T\}$ is simply a set of n jointly distributed random variables.

Stochastic processes having a countably infinite index set are called random walks.

12.1.2 DEFINITIONS: MARKOV PROCESS AND MARKOV CHAIN

In some of the stochastic processes, the "future" depends on the "past" only through the "most recent" observation, let us call this observation the "present" observation.

Markov Process

If the conditional distribution of X_t for any $t > t_0$ (t_0 is viewed as the "present", given $X_{t_1}, X_{t_2}, \ldots, X_{t_n}$, where $t_1 < t_2 < t_3 \ldots t_n, < t_0$, may depend only on the value of the most recent observation X_{t_n}, or

$$P(X_t \angle x/X_{t_1}, X_{t_2}, \ldots, X_{t_n} < t) = P(X_t \angle x/X_{t_n})$$

for all choices of $t_1 < t_2 < t_3 \ldots t_n < t$, then the process

$$\{X_t : t \varepsilon T\}$$

is called a Markov Process

For example, consider a random walk where a particle starts at position 1 at time period zero. At each succeeding time period (1, 2, 3,...), the walk jumps one unit to the right with probability 0.3, and jumps one unit to the left with probability 0.1; it remains stationary with probability 0.6. Assume that the jump at each period is independent of the jumps at other periods. Since the position of the particle at period $t > t_n$ evolves from its position at period *tn*, and it does not depend on how the process managed to get where it was at period t_n, this random walk is called a Markov process.

Definition: Markov Chain

Let $\{X_t : t \varepsilon T\}$ be a Markov process with denumarable parameter space $T = \{0, 1, 2, 3, \ldots\}$, and assume that the random variables in $\{X_t : t \varepsilon T\}$ have as their range a common finite set A of real numbers, called the state space of the process. Then

a. Since the state space of the Markov process $\{X_t : t \varepsilon T\}$ is finite, the process is said to be a finite Markov process.

b. Since T is discrete, the process $\{X_t : t \varepsilon T\}$ is called a finite Markov chain.

12.1.3 STATES AND TRANSITION PROBABILITIES

A Markov chain can be described as follows: there is a set of states $A = \{1, 2, 3, \ldots, n\}$. The process starts in one of these states at time zero and moves successively from one state to another at unit time intervals. Each move is called a step. The probability P_{ij} that the process moves from state i to state j depends only on the state occupied before the step. The probabilities P_{ij}; are called transition probabilities.

Transition Probabilities

Let $\{X_t: T = 0\ 1, 2, 3, \ldots\}$ be a finite Markov chain with state space $A = \{1, 2, 3, \ldots, n\}$. The probability

$$P_{ij}(t) = P(X_t = j / X_{t-1} = i),\ i, j \varepsilon A$$

is called the transition probability from state i to state j. If the $P_{ij}(t)$'s are constant with respect to t, say $P_{ij}(t) = P_{ij}$ the Markov chain is said to be homogeneous, and the P_{ij}'s are called one-step transition probabilities.

Since X_t is a random variable with range A, it follows that $\sum_{j=1}^{n} P_{ij} = 1$, for each i.

12.1.4 TRANSITION MATRIX

The transition probabilities P_{ij} of a finite Markov chain may be arranged in the form of a square matrix called the transition matrix.

Definition: Transition Matrix

A transition matrix $P = \{p_{ij}\}$ is a square matrix of nonnegative entries and row sums equal to 1.

For example, three boys a, b, and c throwing a ball to each other. a always throws the ball to b and b always throws the ball to c; but c is just as likely to throw the ball to b as to a. Let X_{t_n} denote the nth person to be thrown the ball at time t_n. The state space is $A = \{a, b, c\}$.

This is a Markov chain since the person throwing the ball is not influenced by those who previously had the ball. Then the transition matrix of this Markov chain is

$$P = \begin{array}{c} \\ a \\ b \\ c \end{array} \begin{bmatrix} a & b & c \\ 0 & 1 & 0 \\ 0 & 0 & 1 \\ 1/2 & 1/2 & 0 \end{bmatrix}$$

The first row of the matrix corresponds to the fact that a always throws the ball to b. The second row corresponds to the fact that b always throws the ball to c. The last row corresponds to the fact that c throws the ball to a or b with equal probabilities.

12.1.5 *n*-STEP TRANSITION PROBABILITIES

The n-step transition probabilities of homogeneous Markov chains, denoted by $p_{ij}^{(n)}$, are defined as follows:

$$p_{ij}^{(n)} = P(X_{t+n} = j/X_t = i)$$

Let $P^{(n)}$ denote the matrix of the n-step transition probabilities. The following theorem shows the relation between $P^{(n)}$ and P.

Theorem:

If P and P^n are, respectively, the one-step and the n-step transition probabilities of a homogenous Markov chain, with finite state space, then

$$P^n = (P)n$$

12.2 ABSORBING MARKOV CHAINS

12.2.1 DEFINITIONS: absorbing states, and transient states

Absorbing States:

A state i in a Markov chain is called absorbing if it is impossible for the system to leave it once it enters there. That is;

$$p_{ii} = 1$$

Absorbing Chains:

A Markov chain is absorbing if it has at least one absorbing state.

Transient States:

In an absorbing Markov chain, a state which is not absorbing is called a transient state.

Thus, a state i in a Markov chain is absorbing if and only if the ith row of the transition matrix P has a 1 on the main diagonal and zeros everywhere else.

For example, a man walks along a four-block stretch. He starts at corner 1 and, with probability 1/2, walks to the right and, with probability 1/2, walks one block to the left; when he comes to the next corner, he again randomly chooses his direction. He continues until he reaches corner 4, which is a bar, or corner 0, which is a home. If he reaches home or the bar, he stays there.

We form a Markov chain with states 0, 1, 2, 3, and 4. States 0 and 4 are absorbing states. The transition matrix is

$$P = \begin{array}{c} \\ 0 \\ 1 \\ 2 \\ 3 \\ 4 \end{array} \begin{array}{c} \begin{array}{ccccc} 0 & 1 & 2 & 3 & 4 \end{array} \\ \begin{bmatrix} 1 & 0 & 0 & 0 & 0 \\ 1/2 & 0 & 1/2 & 0 & 0 \\ 0 & 1/2 & 0 & 1/2 & 0 \\ 0 & 0 & 1/2 & 0 & 1/2 \\ 0 & 0 & 0 & 0 & 1 \end{bmatrix} \end{array}$$

The states 1, 2, and 3 are transient states and from any of these, it is possible to reach the absorbing state 0 and 4. Hence, the chain is an absorbing chain.

12.2.2 CANONICAL FORM

If in a given absorbing Markov chain, there are r absorbing states, s transient states, and we renumber the states so that the transient states come first, then the transition matrix will have the following canonical or standard form

$$P = \begin{array}{c} \\ \text{Transient} \\ \text{Absorbing} \end{array} \begin{array}{c} \begin{array}{cc} \text{Transient} & \text{Absorbing} \end{array} \\ \left[\begin{array}{c|c} Q & R \\ \hline 0 & I \end{array}\right] \end{array}$$

where, I is an r by r identity matrix, 0 is an r by s zero matrix, R is a nonzero s by r matrix, and Q is an s by s matrix. The first s states are transient and the last r states are absorbing.

12.2.3 PROPERTIES OF ABSORPTION

Recall that the entry $P_{ij}^{(n)}$ of the matrix P^n is the probability of being in the state j, after n steps, when the chain started in state i. $P^{(n)}$ is of the form

$$P^n = \left[\begin{array}{c|c} Q^n & * \\ \hline 0 & I \end{array}\right]$$

where * denotes the s by r matrix in the upper right-hand corner of P^n. Thus, the entries of Q^n give the probabilities for being in each of the transient states after n steps for each possible transient starting state. $Q^0 = I$. When a process reaches an absorbing state, we say that the process is absorbed.

Properties of Absorption

1. In an absorbing Markov chain, the probability that the process will be absorbed is 1. That is;

$$Q^n \rightarrow 0 \text{ as } n \rightarrow \infty$$

That is,

$$\lim_{n \rightarrow \infty} Q^n = 0$$

2. For an absorbing Markov chain, the matrix $I - Q$ has an inverse N, and

$$N = (I - Q)^{-1} = I + Q + Q^2 + Q^3 + \ldots$$

12.2.4 FUNDAMENTAL MATRIX

For an absorbing Markov chain with transition matrix P, the matrix $N = (I - Q)^{-1}$ is called the fundamental matrix for P. The entry p_{ij} of N gives the expected number of times in the transient state j when the process is started in the transient state i.

For example, in the random walk of section 12.2.1, the transition matrix in canonical form is

$$P = \begin{array}{c} \\ 1 \\ 2 \\ 3 \\ 0 \\ 4 \end{array} \begin{array}{c} \begin{array}{ccccc} 1 & 2 & 3 & 0 & 4 \end{array} \\ \left[\begin{array}{ccc|cc} 0 & 1/2 & 0 & 1/2 & 0 \\ 1/2 & 0 & 1/2 & 0 & 0 \\ 0 & 1/2 & 0 & 0 & 1/2 \\ \hline 0 & 0 & 0 & 1 & 0 \\ 0 & 0 & 0 & 0 & 1 \end{array}\right] \end{array}$$

$$Q = \begin{bmatrix} 0 & 1/2 & 0 \\ 1/2 & 0 & 1/2 \\ 0 & 1/2 & 0 \end{bmatrix}$$

and,

$$I - Q = \begin{bmatrix} 1 & -1/2 & 0 \\ -1/2 & 1 & -1/2 \\ 0 & -1/2 & 1 \end{bmatrix}$$

Hence,

$$N = (I - Q)^{-1} \quad \begin{array}{c|ccc} & 1 & 2 & 3 \\ \hline 1 & 3/2 & 1 & 1/2 \\ 2 & 1 & 2 & 1 \\ 3 & 1/2 & 1 & 3/2 \end{array}$$

12.2.5 TIME TO ABSORPTION AND ABSORPTION PROBABILITIES

a. Time To Absorption

Given a Markov chain with s transient states. Let c be an s-component column vector with all entries 1. Then the vector $t = Nc$ has as the jth component the expected number of steps before being absorbed, starting at the state j.

b. Absorption Probabilities

Let b_{ij} be the probability that an absorbing chain will be absorbed in the absorbing state j if it starts at the transient state i. Let B be the matrix with entries b_{ij}. Then B is an s by r matrix, and $B = NR$, where N is the fundamental matrix and R is in the canonical form

In the example of section 12.2.4 , we found that

$$N = \begin{array}{c|ccc} & 1 & 2 & 3 \\ \hline 1 & 3/2 & 1 & 1/2 \\ 2 & 1 & 2 & 1 \\ 3 & 1/2 & 1 & 3/2 \end{array}$$

From the canonical form,

$$R = \begin{array}{c|cc} & 0 & 4 \\ \hline 1 & 1/2 & 0 \\ 2 & 0 & 0 \\ 3 & 0 & 1/2 \end{array}$$

Hence,

$$B = \quad NR = \begin{bmatrix} 3/2 & 1 & 1/2 \\ 1 & 2 & 1 \\ 1/2 & 1 & 3/2 \end{bmatrix} \begin{bmatrix} 1/2 & 0 \\ 0 & 0 \\ 0 & 1/2 \end{bmatrix}$$

$$
= \begin{array}{c} \\ 1 \\ 2 \\ 3 \end{array} \begin{array}{c} \begin{array}{cc} 0 & 4 \end{array} \\ \begin{bmatrix} 3/4 & 1/4 \\ 1/2 & 1/2 \\ 1/4 & 3/4 \end{bmatrix} \end{array}
$$

The first row tells us that, starting from state 1, there is a probability of 3/4 of absorption in state 0, and a probability of 1/4 for absorption in state 4.

12.3 ERGODIC MARKOV CHAINS

12.3.1 REGULAR MARKOV CHAINS

A Markov chain is said to be regular if some power of the transition matrix has only positive elements. That is, there exists an integer m such that all the entries of the transition matrix P^m are all positive.

For example, if the transition matrix of a Markov chain is

$$
P = \begin{bmatrix} 0 & 1 \\ 1/2 & 1/2 \end{bmatrix}
$$

then the chain is regular since

$$
P^2 = \begin{bmatrix} 0 & 1 \\ 1/2 & 1/2 \end{bmatrix} \begin{bmatrix} 0 & 1 \\ 1/2 & 1/2 \end{bmatrix}
$$

But a Markov chain with transition matrix

$$
P = \begin{bmatrix} 1 & 0 \\ 1/2 & 1/2 \end{bmatrix}
$$

is not regular since all powers of P will have a zero in the upper-hand corner.

12.3.2 FIXED VECTORS

A vector $w = (w_1, w_2, \ldots, w_n)$ is said to be a probability vector if all the components are negative and their sum is 1.

If P is a transition matrix for a regular Markov chain, then there exists a unique probability row vector w such that $wP = w$, and any row vector v such that $vP = v$ is a constant multiple of w. Moreover, if c is a constant column vector, then $Pc = c$, and any column vector y such that $P_y = y$ is a constant multiple of c.

> Fixed Vectors
>
> A row vector w with the property $wP = w$ or column vector y such that $Py = y$ is called a fixed vector.
>
> If v is an arbitrary probability vector, then
>
> $$\lim_{n \to \infty} vP^n = w$$

12.3.3 EQUILIBRIUM

> Let P be the transition matrix of a regular Markov chain. Then in the long run, the probability that any state j occurs is approximately equal to the component w of the unique fixed probability vector w of P. That is, the probability of being in the various states after n steps is given by
>
> $$wP^n = w, \text{ as } n \to \infty$$

For example, consider the transition matrix

$$P = \begin{array}{c} \\ a \\ b \\ c \end{array} \begin{array}{c} \begin{array}{ccc} a & b & c \end{array} \\ \begin{bmatrix} 0 & 1 & 0 \\ 0 & 0 & 1 \\ 1/2 & 1/2 & 0 \end{bmatrix} \end{array}$$

of the example presented in section 12.1.4. We seek any fixed vector $t = (x, y, z)$ of P such that

$$(x, y, z) \begin{bmatrix} 0 & 1 & 0 \\ 0 & 0 & 1 \\ 1/2 & 1/2 & 0 \end{bmatrix} = (x, y, z)$$

Thus, $\frac{1}{2}z = x$, $x + \frac{1}{2}z = y$, and $z = y$,. If we let $z = 2$, then we obtain $x = 1$, and $y = 2$. Therefore $t = (1, 2, 2)$ is the fixed vector of P.

But, every multiple of t is a fixed vector of P, hence, let $w = \frac{1}{5}$t, we get the fixed probability vector $w = (\frac{1}{5}, \frac{2}{5}, \frac{2}{5})$. Thus, in the long run, *a* will be thrown the ball 20% of the time, and *b* and *c* 40% of the time.

12.3.4 ERGODIC CHAINS

> Definition:
>
> A Markov chain is said to be an ergodic chain if it is possible to go from every state to every other state.

Obviously, a regular Markov chain is ergodic, because, if the nth power of the transition matrix is positive, it is possible to go from every state to every other state in *n* steps. On the other hand, an ergodic chain is not necessarily regular.

> Theorem
>
> For an ergodic Markov chain, there exists a unique probability vector w such that $wP = w$ and w is strictly positive. Any row vector v such that $vP = v$ is a constant multiple of w. Any column vector y such that $Py = y$ is a constant multiple of column vector c satisfying $Pc = c$.

12.3.5 LAW OF LARGE NUMBERS FOR ERGODIC CHAINS

> Let $H_j^{(n)}$ be the proportion of times in n steps that an ergodic chain is in state j. Then for every $\varepsilon > 0$,
>
> $$P\left(\left|H_j^{(n)} - w_j\right| > \varepsilon\right) \to 0,\ n \to \infty$$
>
> independent of the starting state.

12.3.6 FUNDAMENTAL MATRIX

> Let P be the transition matrix for an ergodic chain. Let the vector $W = (W_1, W_2, \ldots, W_n)$ be the unique probability vector such that $wP = w$; c be column vector with all components equal to l; and an n dimensional $b = (b_1, b_2, \ldots, b_n)$ be any row vector such that $bc \neq 0$. Then the fundamental matrix z determined by b for the given ergodic chain is

$$z = (I - x + cb)^{-1}$$

where I is the identity matrix. This inverse exists, and

$$w = bz$$

For example, let the transition matrix of an ergodic chain be

$$P = \begin{bmatrix} 1/2 & 1/4 & 1/4 \\ 1/2 & 0 & 1/2 \\ 1/4 & 1/4 & 1/2 \end{bmatrix}$$

choosing $b = (1, 1, 1)$, we obtain

$$I - P + cb = \begin{bmatrix} 1 & 0 & 0 \\ 0 & 1 & 0 \\ 0 & 0 & 1 \end{bmatrix} - \begin{bmatrix} 1/2 & 1/4 & 1/4 \\ 1/2 & 0 & 1/2 \\ 1/4 & 1/4 & 1/2 \end{bmatrix} + \begin{bmatrix} 1 & 1 & 1 \\ 1 & 1 & 1 \\ 1 & 1 & 1 \end{bmatrix}$$

$$= \begin{bmatrix} 3/2 & 3/4 & 3/4 \\ 1/2 & 2 & 1/2 \\ 3/4 & 3/4 & 3/2 \end{bmatrix}$$

$$z = (I - P + cb)^{-1} = \begin{bmatrix} 14/15 & -1/5 & -2/5 \\ 2/15 & 3/5 & -2/15 \\ -2/5 & -1/5 & 14/15 \end{bmatrix}$$

and

$$w = (2/5, 1/5, 2/5)$$

12.4 THE FUNDAMENTAL THEOREM FOR REGULAR MARKOV CHAINS

12.4.1 THE AVERAGING THEOREM

If P is an r by r transition matrix for a regular Markov chain, and y is any r-component column vector, then

$$P^n y \rightarrow kc$$

where kc is the constant column vector with components k depending on y. (Remember that c is a column vector with all components equal to 1.)

12.4.2 SMOOTHING LEMMA

Let P be an r by r matrix for a regular Markov chain, and d be the smallest entry of the matrix P. Let y be a column vector with r components, the largest of which is M_0 and the smallest m_0. Let M_1 and m_1 be the largest and smallest component, respectively, of the vector Py. Then

$$M_1 - m_1 \angle (1 - 2d) \times (M_0 - m_0)$$

12.4.3 FUNDAMENTAL LIMIT THEOREM FOR REGULAR MARKOV CHAINS

If P is a transition matrix for a regular Markov chain, then

$$\lim_{n \to \infty} = P^n = cw$$

where w is a probability vector, and c is a column vector with all components equal to 1.

12.5 MEAN FIRST PASSAGE TIME FOR ERGODIC CHAINS

12.5.1 MEAN FIRST PASSAGE TIME

Let P be the transition matrix of an ergodic Markov chain with states $1, 2, 3, \ldots, r$. Let $w = (w_1, w_2, \ldots w_r)$ be the unique probability vector such that $wP = w$. Then, by the law of large numbers for Markov chains, in the long run the process will spend a fraction W_j of the time in state j. Thus, if we start in state i, the chain will eventually reach state j; in fact, it will be in state j in infinitely many times.

Definition: Mean First Passage

If an ergodic Markov chain is started in state i, then the expected number of steps to reach state j for the first time is called the mean first passage time from state i to state j. It is denoted by m_{ij}. By convention, $m_{ii} = 0$.

12.5.2 MEAN RECURRENCE TIME

A quantity that is closely related to the mean first passage time is called the mean recurrence time, defined as follows.

Definition: Mean Recurrence Time

If an egrodic Markov chain is started in state i, then the expected number of steps to return to state i for the first time is called the mean recurrence time for state i. It is denoted by r_i.

It is clear that if we start in state i, we must return to state i, since we either stay at state i the first step or go to some other state j, and from any other state j, we will eventually reach state i.

12.5.3 MEAN FIRST PASSAGE MATRIX AND MEAN RECURRENCE MATRIX

Consider an ergodic Markov chain with m_{ij} the mean first passage time to go from state i to state j, an with r the mean recurrence time to return to state i for the first time.

Mean First Passage Matrix:

Let M be the matrix whose ij entry is the mean first passage time mij. If $i = j$, then the diagonal entries of M are zero. M is called the mean first passage matrix.

Mean Recurrence Matrix:

The matrix D with all entries zero except the diagonal entries $i_{ij} = r_i$ (the mean recurrence time) is called the mean recurrence matrix.